The Modern Homesteader's Handbook:

Sustainable Living in the Digital Age

Learn Step-by-Step How to Raise Crops and Animals in Your Own Backyard

CONTENTS

INTRODUCTION

APPENDIX

CARROT-ZUCCHINI DOUBLE LAYER CAKE

CRÈME DE MENTHE CITRUS SALAD

BLUEBERRY-RASPBERRY STUFFED CREPES

HOME-CANNED MAPLE SPICED APPLESAUCE

NEW ENGLAND CRANBERRY SPICE BREAD

STRAWBERRY GRAHAM CRACKER PIE

BUCKWHEAT-GINGER PANCAKES

MAPLE SPICED GRANOLA

CINNAMON-PUMPKIN MUFFINS

LIGHT AND SAVORY WHEAT BREAD

EGG AND HERB NOODLES

DUCK EGG NUTMEG FRENCH TOAST

DUCK AND APPLE SALAD SANDWICH

POTLUCK BREAKFAST QUICHE

CHICKEN AND ONION ENCHILADA CASSEROLE

BUTTERNUT-GINGER CUSTARD

FARMER'S FRESH GOAT CHEESE

CREAMY GOAT MILK YOGURT

RABBIT SOUP FOR PRESSURE CANNING

SAUSAGE POT PIE

CREAMY CHOCOLATE COOKED PUDDING

BEESWAX AND PEPPERMINT LIP BALM

BEESWAX MOISTURIZING AND HEALING SALVE

HONEY-LEMON CREAM CHEESE SPREAD

SPICED HONEY BUTTER

SPICED HONEY CAKE

REFERENCES

INTRODUCTION

Do you dream of gathering fresh eggs from your own hens? Does saving money and living naturally appeal to you? Transform your backyard into a homestead that puts food on the table, reduces household waste, and helps you return to your roots.

At one time homesteading allowed settlers to move west and claim land by farming it. Today, modern homesteaders choose to raise backyard chickens and tend gardens and orchards to increase sustainability and self-reliance. You, too, can make your backyard a productive oasis in an uncertain world.

No matter why you wish to homestead, you'll find the simple life can be a lot of work. Without a plan to follow and goals to strive for, your path to self-reliance could be rather rocky. Let's dig in and see if you're ready to start.

1. Do you want to reduce waste and increase sustainability?

2. Will you have time to care for your homestead?

3. Are you able to do physical labor?

4. How much space do you have?

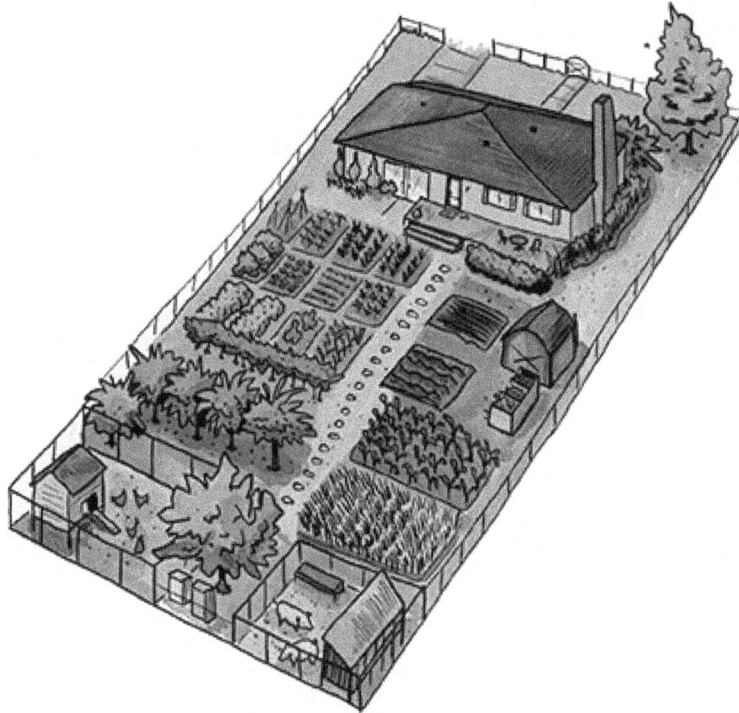

Your responses will shape your homestead plans and projects. With a quarter acre, you could raise small livestock and grow fruits, nuts, vegetables, herbs, and grains. The harvest from your backyard could reduce your grocery bills, fill your pantry, and provide additional income.

Your growing conditions, budget, and time will determine how much you can grow. Assuming you have average conditions and an average lot, in one year you have the potential to raise the following:

- ♦ 6 laying hens to provide 1,200 eggs

- ♦ 25 meat chickens to provide 125 pounds of meat

- ♦ 2 turkeys for the holidays

- ♦ 2 goats to provide 180 gallons of milk

- ♦ 2 beehives to provide 80 to 120 pounds of honey

- ♦ 6 dwarf fruit trees to provide 6 to 24 bushels of fruit

- ♦ 4 nut trees to provide 50 to 150 pounds of nuts

- ♦ Small grains to provide up to 3 bushels of field corn

- ♦ Raised beds to provide 300 to 2,500 pounds of vegetables

To produce these yields, you'll need to purchase livestock feed, plant intensively, and invest in tools and supplies. You can also begin on a smaller scale and increase production as you gain experience.

Throughout this book, you'll find helpful checklists, step-by-step instructions, and budget breakdowns to help you determine which projects to tackle. Beginners are often tempted to try everything immediately, but keep your expectations realistic and enjoy the journey to self-reliance. Let's get started.

CHAPTER ONE

BASIC STEPS FOR PREPARING YOUR HOMESTEAD

Starting a homestead is exciting, but it's important to lay some groundwork before you jump in. This chapter helps you create a plan for your property so you can visualize your projects and determine the most efficient use of your space. You'll also learn about soil type, growing zones, and seasons, all of which affect what you can raise on your homestead. These strategies and tips will help you make the most of what your backyard has to offer, and they will help you customize a plan to fit your needs.

MAP OUT YOUR HOMESTEAD

Planning your homestead on paper can help you prevent costly mistakes. If you map out your existing property features and growing conditions, it will be easier to determine which projects will fit. First, draw your property on graph paper. Check online for an aerial view or request a copy of your plat map, which shows the size and shape of your land, from the county assessor's office. If you can't find this information, measure your lot lines, mark the direction of north, note the scale, and sketch out the following features:

1. House, patio, sidewalk, and driveway

2. Sheds and other outbuildings

3. Existing landscape features (trees, gardens, shrubbery)

4. Property lines

5. Septic field

6. Utility lines (underground and aboveground)

Once your map is complete, lay a sheet of tracing paper on top. Sketch out projects that interest you, taking care to plan around utility lines and septic fields. Make several drawings, moving features around to see how well the layout works.

Your homestead plans may change significantly as you read through the rest of the book. It's better to start out small and add more features later, but the map will give you a much better idea of what will fit on your land.

LEARN MORE ABOUT YOUR SOIL

Before you start planting crops, it's important to learn about your soil, which has a significant impact on crop growth. You need to understand your soil's health, texture, and pH to ensure successful harvests, particularly if you're thinking about investing in picky crops like blueberries and cherries. It's common to find several soil conditions on your property. Note the soil types on your homestead map for future reference.

Soil Texture

To start, dig a hole in each garden area that's 12 inches deep and 12 inches wide. Examine the soil for worm tunnels and rotting plants. Soil that contains signs of life is healthiest for plant growth. There are three main types of soil: clay, silt, and sand. Soil texture affects water drainage and available nutrients. Follow these steps to determine your soil texture:

1. Gather a handful of soil a few inches below the turf.
2. Remove any stones or debris.
3. Add some water and work the soil until it sticks to your hands.
4. Try to roll the soil into a ball. Sandy soil won't form a ball easily, whereas clay rolls into a smooth ball. Press the soil between your fingers and assess whether it is sticky (clay), silky (silt), or gritty (sandy).

Troubleshooting Tip: Sandy soil doesn't hold nutrients or moisture, and heavy clay soil retains water too long. To find out how well your soil drains, fill the hole with water. Allow it to drain overnight, then fill with water again. Use a yardstick to check the water levels every hour. If the water drains at a rate of 1 to 3 inches per hour, most plants will grow well. If the water drains faster or slower, you will need to amend the soil with compost.

SOIL PH

On the pH scale, 0.0 is acid, 14.0 is alkaline, and 7.0 is neutral. Most plants need soil with a pH of 5.5 to 6.5, which is slightly acidic. If your soil has a pH outside this range, you will need to amend it. You can purchase a reusable pH meter or a soil test kit from a garden center or online. The kit should contain tables for lime and sulfur application rates. Alternatively, your local Cooperative Extension Service office will likely perform a pH test for a minimal fee. Expect your pH test to run between $7 and $14, or you can pay a bit more for the reusable meter.

UNDERSTAND YOUR ZONE AND SEASONS

It's important to familiarize yourself with your climate and conditions. Your growing zone and seasons will determine which crops will survive and produce a harvest on your homestead, as well as the best times for seeding and harvesting. You're probably aware that citrus fruits won't survive in northern climates and apple trees won't fruit without cold weather. Additionally, the average date of your last frost in spring will guide your planting dates for a successful harvest.

ZONE

Take a look at the USDA Plant Hardiness Zone Map for more information about your growing zone. It outlines the 26 growing zones in the United States and notes each zone's average minimum temperatures. Use the map to guide your choice of trees, shrubs, and perennials.

When planning for fruit trees, berry bushes, and other perennial plants, look for varieties that thrive in your area. Reputable nurseries list the hardiness zones for each plant they sell. Don't try to cheat a plant's requirements. If you live in zone 4, a tree hardy to zone 5 won't survive a hard winter.

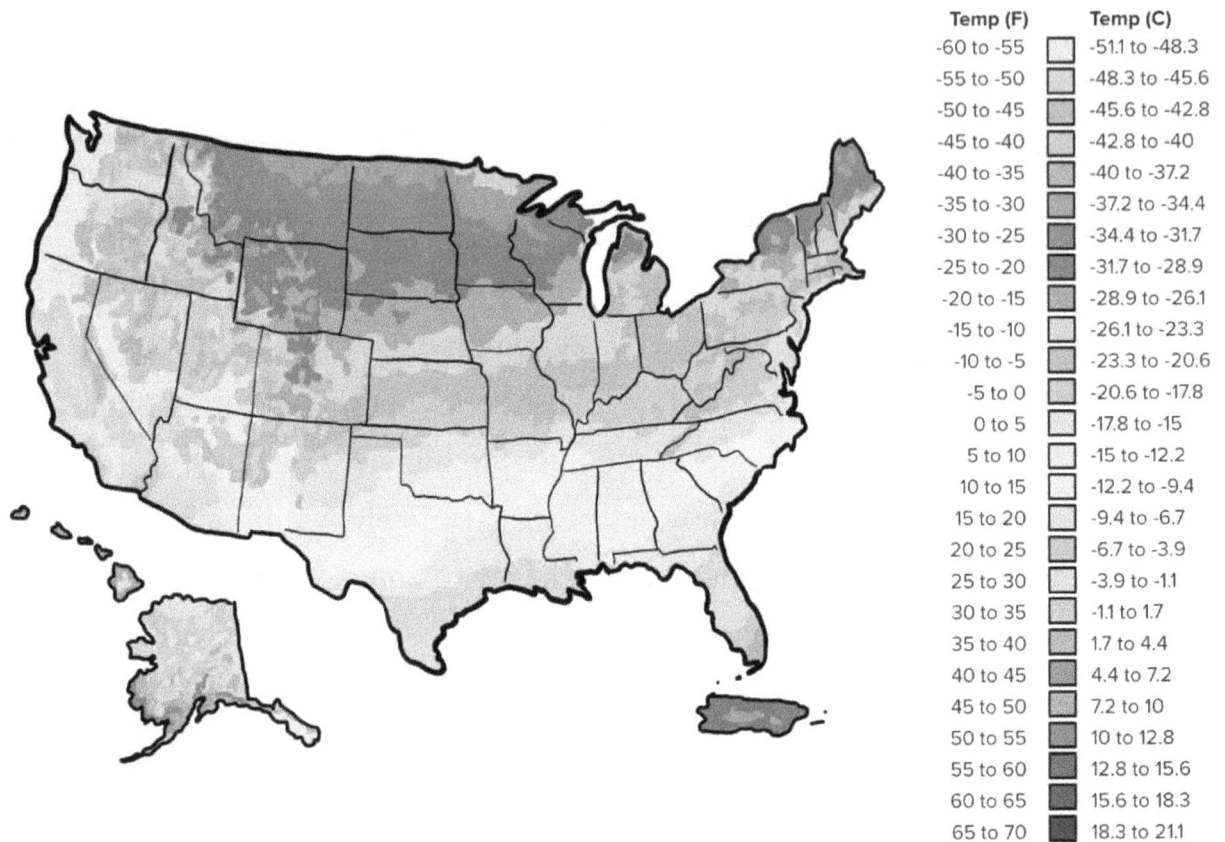

Temp (F)		Temp (C)
-60 to -55		-51.1 to -48.3
-55 to -50		-48.3 to -45.6
-50 to -45		-45.6 to -42.8
-45 to -40		-42.8 to -40
-40 to -35		-40 to -37.2
-35 to -30		-37.2 to -34.4
-30 to -25		-34.4 to -31.7
-25 to -20		-31.7 to -28.9
-20 to -15		-28.9 to -26.1
-15 to -10		-26.1 to -23.3
-10 to -5		-23.3 to -20.6
-5 to 0		-20.6 to -17.8
0 to 5		-17.8 to -15
5 to 10		-15 to -12.2
10 to 15		-12.2 to -9.4
15 to 20		-9.4 to -6.7
20 to 25		-6.7 to -3.9
25 to 30		-3.9 to -1.1
30 to 35		-1.1 to 1.7
35 to 40		1.7 to 4.4
40 to 45		4.4 to 7.2
45 to 50		7.2 to 10
50 to 55		10 to 12.8
55 to 60		12.8 to 15.6
60 to 65		15.6 to 18.3
65 to 70		18.3 to 21.1

SEASONS

It's important to know the average date of your last frost in spring and first frost in autumn. Use an online frost date search, such as freeze/frost data from the National Oceanic and Atmospheric Administration climate normals. This information will help you decide when to plant tender crops and gives you an idea of how long your growing season lasts. Keep in mind these dates are averages, and you could experience frosts later or earlier than those listed.

Check for microclimates, that is, places on your property where conditions are warmer or colder than normal. Tender plants may grow well along south-facing walls, so note these areas on your map. Also mark any low-lying or exposed areas where hardier plants will survive best.

Seasonal changes in temperature and rainfall can be challenging for gardeners. In southern areas, summer heat presents problems, just as cold winters do in the north. Southern gardeners may plant cool-season crops in fall and main crops in late winter to harvest before summer heat sets in. In northern zones, cool-season crops are planted early and main crops go in the ground after the frosts.

Plan to extend your harvests with cold frames in the north and shade in southern summers. Chapter 9 discusses methods for extending your growing season. Check with your local Cooperative Extension Service office for a wealth of information on common pests and diseases, as well as the best crops and varieties for your area.

ASSESS YOUR SPACE

It's exciting to dream about raising your own vegetables, fruits, and laying hens. You probably want to dig right in and grow all of your own food. But just how much can you produce on your homestead?

There's no easy answer to this question. Your projects depend on your local ordinances, growing conditions, time constraints, and budget. But there are a variety of ways to increase the abundance from your land. As you gain experience, you'll find that each year brings new challenges and opportunities on your homestead.

Let's get a general idea of what's possible. Your projects will depend on the nuances of your property. An average lot in the United States is about a quarter acre, and the home, driveway, and patio take up 2,200 to 3,000 square feet. This leaves roughly 8,000 square feet for raising crops and livestock, although you have to account for paths, play areas, and utilities.

With an average backyard on a quarter acre, you have the potential to raise quite a few fruits, vegetables, and herbs, as well as chickens for meat and eggs, bees for honey, goats for milk, grains for flour and livestock feed, and a couple of turkeys for the holidays.

Each chapter gets into more depth concerning the average harvest size and space needed. For the purpose of estimating your potential homestead projects on a sunny quarter-acre lot with decent soil, take a look at our sample harvest plan.

Of course, each property is unique. For example, urban homesteaders might raise bees on a rooftop, keep two or three hens, and grow vegetables and herbs in containers on a balcony. Prioritize the most productive projects and add others as you gain experience. In rural areas, you might raise pigs, scale up your grain plantings, and tend a market garden. Read through the rest of the book to get a feel for what you can grow.

Here are some things to keep in mind as you estimate the size of your potential harvest:

Local ordinances. Check these before you begin. Some areas allow chickens and gardens but prohibit larger animals, even on rural properties. Activities such as processing meat animals or composting may be restricted.

Your property's limitations. Shade, lack of irrigation, poor soil, and short seasons can impact your growing space. Keep in mind that growing intensively creates the potential for disease and pest problems. For each action plan you create, take the time to research potential issues and plan for treatments.

Labor. Be mindful of the labor necessary for each homesteading project. Take advantage of time-saving methods in the garden, such as mulching and drip irrigation. Remember that livestock require daily care. Feeding, cleaning stalls, and milking regularly is crucial for the health of the animals.

Financial investment. Almost every project you undertake will also require an initial cash investment. A few chicks are cheap, but the cost of caring for them adds up. And, of course, the larger the livestock, the more food they consume. It can be expensive to invest in livestock, pay vet bills, and provide grain, hay, and bedding.

Read through the chapter on each homestead feature for cost estimates and payback on your investment. Chapter 2 walks you through making realistic plans to avoid the pitfalls of homesteading.

SAMPLE BACKYARD HARVEST PLAN (SUNNY QUARTER-ACRE LOT)

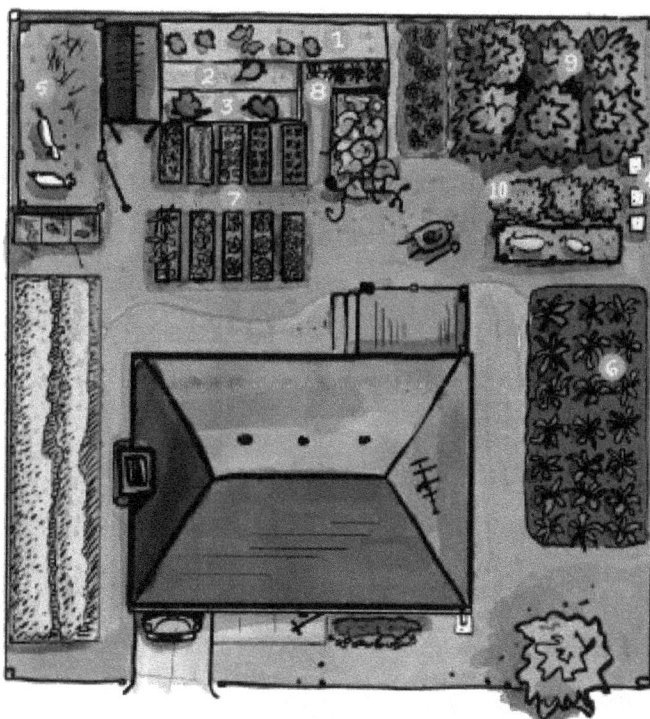

LEGEND

1. **6 TO 12 EGG-LAYING HENS**, 42 to 120 sq. ft. minimum, 1,200 to 3,600 eggs per year.
2. **25 MEAT CHICKENS**, 50 sq. ft. minimum, 125 pounds of meat in 7 to 8 weeks.
3. **2 TURKEYS**, 10 sq. ft. minimum, 30 to 50 pounds of meat in 5 to 6 months.
4. **2 BEEHIVES**, 4 sq. ft. (plus room to access the hive), 80 to 120 pounds of honey per year.

5.   2 NIGERIAN DWARF MILK GOATS, 70 sq. ft., 180 gallons of milk per year.
6.   FIELD CORN, 800 sq. ft., 2 to 3 bushels per year.
7.   VEGETABLES, 480 to 960 sq. ft., 300 to 2,500 pounds per year.
8.   HERBS, 40 to 80 sq. ft., 10 to 40 pounds per year.
9.   4 TO 6 DWARF FRUIT TREES, 200 to 600 sq. ft., 8 to 24 bushels per year.
10. 2 TO 4 SMALL NUT TREES, 200 to 1,200 sq. ft., 25 to 80 pounds per year.

MAXIMIZE YOUR SPACE

With careful planning and efficient use of space, you can raise a great deal of food from your homestead. Each chapter helps you choose the crops and animals that produce the best harvests for your space and needs. Maximize harvests on your homestead with these techniques:

Use raised beds. Growing vegetables, herbs, and small fruits in raised beds allows you to get a jump start in spring and raise crops in areas with poor drainage. Learn more in chapter 3.

Use square-foot gardening. Grow up to twice the harvest with a square-foot garden instead of a traditional row garden.

Use succession planting. Sow seeds in succession to harvest up to three times the amount of food from one bed. Early greens may be harvested in a month or so. As you harvest spring salads, replace these crops with peppers, tomatoes, or basil. As fall draws near, plant another round of cool-season vegetables. Learn more in chapter 3.

Grow vertically and in containers. Use a trellis to grow vines, such as cucumbers, in less than half the space. Container growing allows you to raise food on a patio or even a balcony.

Improve your soil with composting. Set up a composting system to create rich, black soil for improved soil health and plant growth. With increased nutrients in the soil, you can plant vegetables intensively to crowd out weeds.

Consider raising small livestock. Chickens, rabbits, dwarf goats, and bees may be great choices for homesteaders in a small space. A variety of small livestock can increase your land's production with milk, eggs, honey, and meat while increasing soil fertility with composted manure. Learn more in chapter 6, chapter 7, and chapter 8.

As you develop your homestead plans, allow space for rotating crops and pastures to reduce parasites, disease, and pests. After you harvest, allow chickens or pigs to forage for pests while turning and fertilizing the soil.

PLAN TO DIVERSIFY

Whether you're looking to increase self-reliance or make a side income from your homestead, it's crucial to diversify your harvests. This ensures you'll have food or income even if one project fails.

As you choose your crops, livestock, and products, think about the seasons in which they'll produce and how long the harvest will last. Plan for production over an extended period of time. Consider how to extend your season or preserve extras for later. Select plants and livestock that supply food or sales all year.

Here are some common crops and livestock and when they produce food in a temperate growing zone:

Cool-season vegetables. Peas, lettuce, broccoli, and other crops are harvested in spring and fall. Use cold frames to extend the season into winter.

Warm-season vegetables. Tomatoes, beans, sweet corn, and many other crops are harvested in summer and fall. Plant after all chance of frost.

Herbs. Plant basil, savory, and sage in the garden for summer harvest or indoors in containers for winter use.

Grains. Oats, wheat, and field corn are harvested in summer and fall and store well for winter use.

Roots and storage crops. Plant carrots, potatoes, winter squash, and other good keepers for fall harvest and winter storage in a root cellar.

Poultry for meat and eggs. Hens will lay eggs all year with supplemental light in winter. Harvest old hens and raise broilers for meat all year.

Goats. You can breed two goats at different times for year-round milk production.

Rabbits. A breeding trio of rabbits provides meat all year.

Fruits and nuts. Raise these for summer and fall harvests and preserve for winter.

Spring through fall are the most productive seasons on a homestead, but you can extend your seasons and preserve the harvest to last all year with planning.

CHAPTER TWO

HOW TO MANAGE YOUR HOMESTEAD

The skills necessary for operating a homestead business aren't second nature for most beginners. Your success depends on thoughtful planning, experience, and following through on your goals. An action plan provides a crucial road map for completing tasks and projects. Setting weekly, monthly, and yearly goals will guide you along the path to self-reliance.

Are you dreaming about gardening and maybe raising a few chickens to increase your self-sufficiency? Perhaps you'd like to start a side business to supplement your income. In either case, you need to have a plan in place to achieve SMART goals, or those that are Specific, Measurable, Attainable, Realistic, and Timely.

This chapter helps you create an action plan, set goals, and break down the goals into manageable steps. As you work through your plans, think about the skills required to achieve your goals. Do you know how to garden, raise livestock, and preserve food? This book provides step-by-step instructions for raising specific crops and livestock; maximizing, extending, and preserving harvests; and generating a side income from your homestead. You'll find recipes for using your bounty in the appendix.

CREATE A PROJECT ACTION PLAN

Whether you live on a suburban lot or 15 acres, it's essential to start with an action plan to guide your vision. Begin by answering a few questions to build the foundation of your plan, then record your answers on the Homestead Action Plan template.

- Do you want to homestead for business reasons, personal reasons, or both?

- Will you raise livestock and gardens for your own use?

- Do you intend to earn an income from your homestead?

- What is the core identity of your homestead? What projects align with your family's values? What do you hope to achieve on your homestead?

  ↗ GROW ORGANIC FRUITS, VEGETABLES, AND GRAINS

  ↗ PROVIDE HABITAT FOR POLLINATORS, BIRDS, AND OTHER WILDLIFE

  ↗ LIVE A MORE SUSTAINABLE LIFESTYLE

  ↗ GROW HEALTHY FOOD FOR SALE

- Why do you want to live the simple life? Is there an underlying theme that's the *core identity* of your homestead?

Use the answers to create a homestead mission statement, two or three sentences that will guide your goals and plans and keep you true to your values. (See the Homestead Action Plan.) Here is an example:

The mission of my homestead is to provide organic fruits and vegetables and humanely raised eggs for my family and for sale at our local farmers' market. Our homestead will allow us to reduce our waste, increase our self-reliance, and earn additional income.

SETTING YOUR HOMESTEAD AND HARVEST GOALS

Now take some time to record your general goals for your homestead. Use this example as a guideline for creating a calendar that lays out the process of purchasing materials, constructing livestock housing, planting crops, or marketing a homestead business. The example below sets weekly goals for chicken tasks.

Week 1. Research laying hen breeds and choose a chicken coop design for 25 chickens. Source materials for the chicken coop and estimate the cost.

Week 2. Purchase materials to build a chicken coop. Map out 15-by-25-foot vegetable garden and prepare the soil for planting.

Week 3. Frame in the chicken coop and build the roof. Plant kale, collards, lettuce, and peas.

Week 4. Finish siding the chicken coop. Purchase 15 fence posts and a 100-foot roll of chicken wire. Order 25 chicks and purchase a heat lamp, feeder, and waterer, and set up a brooder for chicks. Start tomatoes, peppers, and eggplants under lights indoors.

List your short- and long-term goals and projects in a notebook or journal, and continue to update your plans as you work through this book.

PLANNING FOR SPECIFIC HARVESTS

Your Homestead Action Plan is a general guide for beginning your backyard farming adventures and is a fantastic first step toward self-reliance. The following chapters delve into each aspect of homesteading, from gardening and raising livestock to preserving harvests. Each section of the action plan includes the steps to create an action plan for your customized road map to homesteading success. Here are the basics:

1. List the overall goals for your homestead.

2. Break these down into specific project goals for your first season.

3. Research the cost, necessary space and time, and potential problems for each project.

4. Determine whether each goal is feasible or if you need to modify your projects.

5. Break down the estimated costs as well as the weekly and monthly tasks required to reach your goals.

KEEP YOUR GOALS REALISTIC

Although you have the potential to raise a lot of food in your backyard, you also have to invest time, energy, and money. Every plant and animal has requirements and limitations. As you set each goal for your homestead, keep these realities in mind.

Perhaps one of your goals is to raise chickens. Before you purchase chicks, you need to determine the feasibility of keeping chickens, create an action plan and set your goals. The following example works through adding laying hens to the homestead.

SAMPLE FEASIBILITY CHECKLIST

- ↗ HOW MUCH SPACE DO YOU HAVE TO RAISE CHICKENS? *Provide each chicken with at least 7 square feet of room in the coop and pen. More is better.*

- ↗ WHAT IS YOUR YEARLY BUDGET FOR KEEPING CHICKENS? *You will spend $25+ per year to feed each hen. Additional costs are covered in more detail in* chapter 6.

- ↗ HOW MANY EGGS DO YOU USE EACH WEEK? *Keep track of your egg consumption to determine the size of your flock.*

- ↗ WHAT ARE THE POTENTIAL PROBLEMS INVOLVED IN KEEPING CHICKENS AND HOW COULD YOU SOLVE THOSE ISSUES? *Prepare for predators and disease.*

Sample Homestead Goal

Add a flock of six laying hens to our homestead this year to provide 30+ eggs each week.

Sample Action Plan

In chapter 6, you'll find cost estimates for the various materials listed here. Check the availability and price of materials in your area when doing your own estimates.

Sample First-Year Goals and Plans

Using the Breakdown of Yearly Goals template, fill in as many details as you can for a practical cost and time estimate. Keep in mind that you may spend more money and expend more effort than anticipated. Some projects take less time, so use the Breakdown of Weekly Goals or the Breakdown of Monthly Goals when applicable.

MONTH	GOALS TO ACCOMPLISH AND MATERIALS NEEDED
JAN	⬈ Research laying breeds, cost of supplies, and local regulations for egg sales. **Time: 2 hours** ⬈ Track how many eggs we use each week. **Time: 15 minutes per week**
FEB	⬈ Determine the cost to convert the garden shed to a coop. **Time: 1 hour** ⬈ Determine the size of flock and space needed. **Time: 30 minutes**
MAR	⬈ WEEK 1: Convert the garden shed into a chicken coop and add a pasture. (Materials needed: 8 ft. sapling for roost, 50 ft. of chicken wire ($30), 10 fence posts ($3.50 each). **Total cost: $65 / Time: 4 to 5 hours** ⬈ WEEK 2: Purchase a chick feeder and waterer ($12) and heat lamp and bulb ($11). Set up brooder. **Total cost: $23 / Time: 1 hour** ⬈ WEEK 3: Purchase the chick starter feed ($13.50 for 50 lbs.), chick grit ($9), probiotics ($8), pine shavings ($5), and 6 hybrid layer pullet chicks ($4 each). **Total cost: $59.50 / Time: 1 to 2 hours** ⬈ Plan on 30 minutes of daily care × 14 days. **Time: 7 hours** ⬈ WEEKS 3 TO 6: Monitor chick health brooder temperature during daily care routine.
APR	⬈ WEEK 5: Set up a brooder area in the chicken coop. **Time: 1 hour** ⬈ WEEK 6: Move pullets to their new brooder area if conditions allow. **Time: 30 minutes** ⬈ WEEK 7: Allow the pullets outside if weather allows. **Time: 15 minutes** ⬈ WEEK 8: Turn the heat lamp off if weather allows. ⬈ Plan on 30 minutes of daily care × 30 days. **Time: 15 hours**
MAY	⬈ WEEKS 9 TO 12: About 30 minutes of daily care. **Time: 15.5 hours** ⬈ Purchase the feed and bedding as needed. **Cost: $18.50 / Time: 30 minutes**
JUN	⬈ WEEKS 13 TO 17: About 30 minutes of daily care. **Time: 15 hours** ⬈ Purchase the feed and bedding as needed. **Cost: $18.50 / Time: 30 minutes**
JUL	⬈ WEEKS 18 TO 22: About 30 minutes of daily care. **Time: 15.5 hours**
AUG	⬈ WEEKS 23 TO 27: About 45 minutes of daily care. **Time: 23.25 hours** ⬈ Hybrid pullets should begin laying soon. Check for eggs during daily care routine.

↗ Purchase 50 lbs. of layer feed and bedding as needed. **Total cost: $17.50 / Time: 30 minutes**

SEP

↗ WEEKS 28 TO 32: About 45 minutes of daily care. **Time: 22.5 hours**

↗ Purchase 50 lbs. of layer feed. **Total cost: $12.50 / Time: 30 minutes**

↗ Sell 18 eggs per week for $5/dozen. **Total gain: $30 / Time: 15 minutes**

↗ Buy 10 bales of straw for winter. **Total cost: $50 / Time: 1 hour**

↗ Put the light on a timer for 14 hours a day to encourage egg production. **Total cost: $15 / Time: 15 minutes**

OCT

↗ WEEKS 33 TO 37: About 45 minutes of daily care. **Time: 23.25 hours**

↗ Purchase 50 lbs. of layer feed. **Total cost: $12.50 / Time: 30 minutes**

↗ Sell 18 eggs per week for $5/dozen. **Total gain: $30 / Time: 15 minutes**

↗ Purchase a heater for the water container for winter. **Total cost: $30 / Time: 15 minutes**

NOV

↗ WEEKS 38 TO 42: About 45 minutes of daily care. **Time: 22.5 hours**

↗ Prepare the chicken coop for winter. **Time: 2 hours**

↗ Clean the coop and set up water heater. **Time: 30 minutes**

↗ Purchase 50 lbs. of layer feed. **Total cost: $12.50 / Time: 30 minutes**

↗ Sell 18 eggs per week for $5/dozen. **Total gain: $30 / Time: 15 minutes**

DEC

↗ WEEKS 43 TO 47: About 45 minutes of daily care. **Time: 23.25 hours**

↗ Purchase 50 lbs. of layer feed. **Total cost: $12.50 / Time: 30 minutes**

↗ Sell 18 eggs per week for $5/dozen. **Total gain: $30 / Time: 15 minutes**

Total Investment: Total cost: $347 / Time: 87.75+ hours
Total Value of Eggs Produced: Total gain: $120
Net Expenses after Egg Sales and Savings: Total cost: $227

As you read through this book, think about the challenges you'll face. Be realistic about your budget and available time to care for all your planned crops and livestock.

SET UP A HOMESTEAD CALENDAR

You will always have projects to keep you busy on a homestead. Plan out your year as completely as possible. This will provide the structure you need to complete chores at the best time. Some chores are daily obligations, and others are seasonal.

You can use a paper calendar or an online one that sends reminders to your phone. Schedule tasks on your homestead calendar to stay on track. Here is an example of a schedule for raising hybrid broiler chickens.

WEEK	BROILER CHICKEN CALENDAR: 8 WEEKS FROM HATCH TO PROCESSING
	Before the chicks arrive, prepare a brooder box and purchase feed and bedding.
1	Keep the brooder temperature at 95°F. Purchase the chicks and provide feed, water, and clean bedding daily. Remove the feed overnight throughout the entire 8 weeks to prevent health issues.
2	Keep the brooder temperature at 90°F. Provide feed, water, and clean bedding daily.
3	Keep the brooder temperature at 85°F. Provide feed, water, and clean bedding daily. Add extra feed and water containers as needed.
4	Keep the brooder temperature at 80°F. Provide feed, water, and clean bedding daily.
5	Keep the brooder temperature at 75°F. Provide feed, water, and clean bedding daily.
6	Move meat chickens to a chicken tractor if weather allows. Provide feed and water daily. Move the tractor to fresh pasture.
7	Provide feed and water daily. Move the tractor to fresh pasture.
8	Provide feed and water daily. Move the tractor to fresh pasture. The day before processing, remove feed 24 hours before processing, and provide clean water.

MAKE A LIST OF HOMESTEAD CHORES

Every homestead will have different daily and seasonal chores. Be sure to allow time in your calendar for daily responsibilities, such as feeding and watering livestock, cleaning stalls, tending to household chores, and tracking expenses. Note the amount of time spent on projects so you'll have a better idea of whether you have the capacity to add new ones. Here are some seasonal homestead chores:

Spring

↗ START SEEDLINGS

↗ TILL THE GARDEN AND GRAIN FIELDS

↗ TRANSPLANT THE SEEDLINGS

↗ GROW COOL-SEASON CROPS IN COLD FRAMES

↗ SOW GRAINS AND VEGETABLES

↗ ORDER AND CARE FOR CHICKS

↗ ASSIST LIVESTOCK BIRTHS

↗ MILK DAIRY ANIMALS

↗ MAKE CHEESE AND DAIRY PRODUCTS

↗ SET UP NEW BEEHIVES

↗ FERTILIZE FRUIT AND NUT TREES

↗ HARVEST EARLY-SEASON CROPS

↗ WEED, MULCH, COMPOST, AND AMEND SOIL

↗ SPRAY FRUIT CROPS

↗ CLEAN OUT ROOT CELLAR

Summer

- ↗ WEED, MULCH, AND COMPOST
- ↗ WATER THE CROPS
- ↗ HARVEST THE CROPS
- ↗ PREPARE THE BEDS FOR FALL CROPS
- ↗ PRESERVE FRUITS, VEGETABLES, AND HERBS FOR WINTER
- ↗ BUTCHER EARLY MEAT HARVEST (CHICKENS, RABBITS)
- ↗ MILK DAIRY ANIMALS
- ↗ MAKE CHEESE AND DAIRY PRODUCTS
- ↗ SPRAY FRUIT CROPS
- ↗ REMOVE EXCESS FRUIT FROM TREES
- ↗ HARVEST HONEY
- ↗ CHECK THE HIVES AND ADD SUPERS AS NEEDED
- ↗ CATCH SWARMS WHEN AVAILABLE
- ↗ CHECK THE HIVES FOR PESTS AND DISEASE

Fall

- ↗ HARVEST AND PRESERVE CROPS; STORE ROOT CROPS
- ↗ BUTCHER LIVESTOCK
- ↗ STORE FEED, HAY, AND BEDDING FOR WINTER
- ↗ BREED SOME LIVESTOCK FOR SPRING BIRTHING

- ↗ TAKE STOCK OF HARVESTS AND PLAN FOR THE COMING YEAR
- ↗ SELL EXCESS LIVESTOCK
- ↗ CLEAN CHIMNEY AND PUT UP FIREWOOD
- ↗ CLEAN DEBRIS FROM GARDENS AND ORCHARDS

Winter

- ↗ CARE FOR LIVESTOCK
- ↗ BREED LIVESTOCK
- ↗ BUTCHER LIVESTOCK
- ↗ CHECK ON THE BEES AND FEED THEM WHEN NECESSARY
- ↗ PLAN THE VEGETABLE GARDEN AND GRAINS
- ↗ PURCHASE SEEDS AND MATERIALS
- ↗ PRUNE FRUIT TREES (LATE WINTER)
- ↗ APPLY DORMANT OIL SPRAY TO FRUIT AND NUT TREES (LATE WINTER)
- ↗ BUILD BEEHIVES AND SUPERS
- ↗ REPAIR FENCES AS NEEDED
- ↗ SPLIT FIREWOOD
- ↗ TOTAL EXPENSES AND PLAN NEXT YEAR'S BUDGET

Seasonal Homesteading Tip: Use your downtime in winter to research new projects for the coming year. Learn about the health needs of livestock or best growing conditions for crops, and record this information for easy reference.

MAP OUT YOUR HARVESTS

Once you determine which crops and livestock you want to raise, add them to your homestead map. Use a sheet of tracing paper to draw vegetable beds, herb gardens, fruit and nut plantings, grain plantings, the chicken coop, the goat shed, animal pens, and the beehives. Rework these features until you are satisfied with their locations. Keep in mind the availability of water, electricity for outbuildings, and the location of utilities. Consider this a work in progress as you gain experience and try new projects.

Once you establish the basic layout of your homestead, list the square footage for each feature, note what you plan to raise in that space, and create a log to track future production. Your log should include the specific project, expenses, notes, and yield. In your notes, include information like seasonal conditions, problems encountered, and changes to make in the future. Does your vegetable garden need more fertilizer or should you keep a more productive breed of chicken? These observations are your key to success.

Each time you harvest food from your homestead, record the yield in your log (see template below). Tracking these harvests and the cost to produce them will help you decide which crops and livestock give you a reasonable return on your investment.

Tips for Tracking Yields: Keep a high-quality scale in your kitchen for weighing vegetables, herbs, and meat. Weigh produce before you wash it and meat right after processing it. Then, record the yield. Some harvests are easier to track using bushels, pints, or other units of measurement. Keep a running tally and enter the total harvests in your weekly production log.

TOOLS FOR MANAGING YOUR HOMESTEAD

Use the following tools as you develop resources and plans for your homestead.

Harvest Log for Crops or Livestock

CROP OR LIVESTOCK: _____ YEAR: _____

GENERAL NOTES: (e.g., record low rainfall, late frosts, or other seasonal factors that affect crops, as well as predator attacks or disease that affect livestock production)

WEEK	COST	LOCATION	PRODUCTION NOTES	YIELD

Using the example of raising laying hens, here's how to use your harvest log:

- ↗ Record the number of eggs collected from your hens each day.
- ↗ Track the income produced from the sale of eggs, chicks, or laying hens.
- ↗ List the problems encountered, the treatments used, and if the remedy was successful.

↗ Record all expenditures, including cost of chicks, feed, bedding, medications, vet bills, supplies, and electricity.

↗ At the end of the year, total expenses, yields, income, and losses. Use this to determine the cost to produce eggs for the year.

↗ If losses are high or expenses outweigh the savings or income, look at your notes to see what went wrong and determine how you may address those problems in the future.

Homestead Action Plan

Use this template to record your action plan and create a mission statement for your homestead.

HOMESTEAD NAME: _____

Do you intend to homestead for personal or business purposes?

↗ Personal

↗ Side income

↗ Full-time income

1. Why do you wish to homestead? List every reason that comes to mind.

2. List the crops and livestock you'd like to raise on your homestead. Do you want fruits, vegetables, herbs, grains, chickens, dairy animals, bees, or some combination thereof?

3. List the services you wish to provide for a side income. Will you provide landscaping, animal boarding, or carpentry work?

4. How will you diversify your homestead harvests? Brainstorm crops or products to sustain your family or to sell for profit in every season.

5. List the personal values that will guide your homestead plans. Is stewardship of the earth, self-reliance, or spirituality a priority?

6. What is the core identity of your homestead? This is the direction in which your homestead will grow and how it is guided by your values.

7. Create a mission statement for your homestead. Use the answers to previous questions to write a few sentences describing your vision for the future.

8. List your goals for the first year on your homestead.

HOMESTEAD PROJECT ACTION PLAN (HPAP) TEMPLATE

Use this template to record action plans for each crop or livestock project on your homestead. As you work through the rest of the book and learn the individual considerations of each project, refer back to this template to help guide your project-specific action plans. We've included abbreviated tables to get you started.

HOMESTEAD PROJECT: _____ YEAR: _____

DESCRIPTION OF PROJECT: _____

ATTACH MAP OF PROJECT (AS NEEDED)

FEASIBILITY RESEARCH

Check into the local cost of materials, the amount of time needed to complete tasks, and the potential for increasing your self-reliance or generating a side income (e.g., what price will your local market support?) to determine whether you have the budget and time to take on this project. Complete the following tables for a basic understanding of the needs for each project.

Expected Cost of Materials

List all materials needed to set up and maintain the project to completion or for one year.

MATERIALS	ONE-TIME VS. REGULAR EXPENSE	COST ESTIMATE

TOTAL COST EXPECTED FOR PROJECT OR FOR FIRST YEAR: _____

EXPECTED TIME TO YIELD: _____

Expected Yield Per Year

How much food do you estimate this project will yield in a season? Use this table to develop a better idea of how much you can gain from each crop or livestock raised on your homestead.

HOMESTEAD PROJECT	NUMBER RAISED	EXPECTED YIELD

Estimated Time

Record the expected time commitment for your potential project. Look at the initial setup and daily tasks, and jot these down. Compare your available time with these estimates to determine whether the project is feasible. Choose the projects that best fit your lifestyle.

Potential Problems and Cost of Solutions

Take time to research pests, diseases, and other problems you may encounter for each project. Check with your local Cooperative Extension Service office for common issues in your area. How can you identify, avoid, and treat these problems? Note whether you need to buy any medications or natural pest control.

IS THE PROJECT FEASIBLE THIS YEAR? _____

After researching the time commitment, costs, and potential problems,
do you feel the project is feasible this year? Is there a way to scale
back or modify your plans to fit within the budget or time frame
available?

PROJECT ACTION PLAN

Look at the list of homestead projects from your information sheet and
break them down into weekly and monthly goals. For example, if you
wish to add gardens and fruit trees this year, jot down the plants you
want and the necessary steps to add them. When should you plant
trees and prepare your garden space? Use the following weekly,
monthly, and first-year goal tables to plan your projects in more detail.

Breakdown of WEEKLY Goals

HOMESTEAD PROJECT: _____ WEEK:

_____

List your daily tasks, expected time to complete, and cost of materials
needed to reach your weekly goals.

DAY	TASKS TO COMPLETE AND MATERIALS NEEDED	TIME	COST
MONDAY DATE: _____			
TUESDAY			

DATE:				
	TOTALS			

Breakdown of MONTHLY Goals

HOMESTEAD PROJECT: _____ MONTH/YEAR:

_____

List the weekly tasks to accomplish, materials needed, time, and cost necessary to reach your monthly goals.

WEEK	GOALS TO ACCOMPLISH AND MATERIALS NEEDED	TIME	COST
WEEK 1 DATES: _____			
WEEK 2 DATES: _____			
	TOTALS		

Breakdown of YEARLY Goals

HOMESTEAD PROJECT: _____ YEAR:

_____

List the monthly tasks to accomplish, materials needed, time, and cost necessary to reach your yearly goals.

MONTH	GOALS TO ACCOMPLISH AND MATERIALS NEEDED	TIME	COST

JAN			
FEB			
	TOTALS		

CHAPTER THREE

VEGETABLE AND HERB GARDENS

S tarting a garden is a cost-effective way to increase food security, but you need more than a handful of seeds to grow your own vegetables and herbs. With a clear plan, reasonable goals, and determination, you can harvest enough food to significantly reduce your grocery bill.

Successful gardeners raise crops best suited to their growing conditions. This chapter helps you determine which vegetables and herbs make the most sense for your homestead.

You'll also create an action plan for the first growing season. We'll discuss how to rotate crops, prepare your soil, build raised beds, plant in succession, and maintain a healthy garden. Check out the cost estimates, tables for choosing the best vegetables and herbs, and seasonal chores checklist. If you're interested in preparing fresh dishes with your garden's bounty, see the recipes here.

CREATE A PROJECT ACTION PLAN FOR YOUR VEGETABLE AND HERB GARDEN

Each time you pluck tomatoes or basil from the garden, you're getting the freshest produce possible. With planning and regular care, a vegetable garden will provide an abundance of inexpensive food.

Use the Homestead Project Action Plan (HPAP) template to flesh out an action plan for your vegetable and herb garden. The following sections review all the considerations involved in creating a successful garden.

FEASIBILITY RESEARCH

To determine the feasibility of your garden project, let's compare the cost of purchasing produce with raising your own and estimate the time commitment.

The USDA's Economic Research Service (ERS) 2018 report on the retail price of fruits and vegetables lists the average costs per pound and serving for a variety of produce. According to the report, it was possible to purchase the daily vegetable requirement for a 2,000-calorie diet (2½ cups) for approximately $0.88. Here's the report's average retail prices for some common vegetables:

- 1 cup of fresh potatoes: $0.20
- 1 cup of frozen green beans: $0.56
- ½ cup of fresh, raw carrots: $0.12

To calculate the annual cost of vegetables needed per person, multiply the daily cost by 365: $0.88 × 365 = $321.20.

Your costs will vary according to local prices. When comparing vegetable costs, be sure to check the size of the package. Frozen vegetables are often sold in 12-ounce bags, so calculate the cost per ounce and multiply that by 16 to find the cost per pound.

The following table compares the average retail prices of common vegetables with the cost of seeds to grow your own. If you have limited space, concentrate on growing the foods you like best that also offer the most savings. For example, the table shows that large tomatoes offer the greatest cost savings.

This table assumes you are planting seeds, with the exception of potato sets. The average cost to plant per square foot is based on typical plant spacing (see the Summer and Winter Vegetables and Herbs table) and typical seed costs in 2020.

POTENTIAL SAVINGS FOR COMMON HOMEGROWN VEGETABLES

VEGETABLE	AVERAGE RETAIL COST/LB.	HOW MANY LBS. IN A 1-CUP SERVING	AVERAGE COST TO PLANT/SQ. FT.	YIELD IN LBS./SQ. FT.	COST TO PRODUCE 1 LB.	SAVINGS/LB
BEANS, GREEN	$1.70 (frozen)	0.298	$0.26	0.2	$1.30	$0.40
CARROTS	$0.77 (fresh)	0.276	$0.47	1.0	$0.47	$0.30
ONIONS, STORAGE	$1.05 (fresh)	0.353	$0.26	1.3	$0.20	$0.85
POTATOES	$0.60 (fresh)	0.265	$0.30	1.0	$0.30	$0.30

SWEET CORN	$1.60 (frozen)	0.364	$0.03	0.3	$0.10	$1.50
TOMATOES, LARGE ROUND	$2.01 (fresh)	0.375	$0.07	2.0	$0.04	$1.97
ZUCCHINI	$1.64 (fresh)	0.397	$0.02	0.75	$0.03	$1.61

Retail prices from USDA Economic Research Service Fruit and Vegetable Prices; cost to grow based on typical seed and potato set prices in March 2020; yields based on USDA Vegetables 2019 Summary.

To calculate savings for each crop, use the Harvest Log template to track expenses and yield per square foot to determine your cost per pound to raise each vegetable. Use this price to compare your production costs with the retail price. Here's how to calculate your final cost per pound for each crop: *cost to plant / yield in pounds = cost to produce per pound.*

In your action plan, determine whether you have the budget and time to raise vegetables and herbs. Compare the expected cost of materials to local prices and determine your budget. Read through the seasonal chores checklist for the approximate time commitment for garden tasks.

EXPECTED COST OF MATERIALS

Your action plan should include a section for estimating basic costs. List all the materials you need for the first year, including seeds, plants, pots, seed starting materials, tools, fertilizers, and gloves. Note one-time expenditures, too.*

Basic Vegetable and Herb Garden Cost Estimates:

- Seeds: $1 to $2 per pack

- Potting soil: $15 for 2 cubic feet

- ◆ Seed starting trays: $10 for 72 cell trays

- ◆ Garden spade*: $10 to $35

- ◆ Hoe*: $10 to $25

- ◆ Garden rake*: $10 to $25

- ◆ Trowel*: $2 to $12

- ◆ Tomato cages*: $2 to $20 each

Based on typical prices found online in March 2020.

*You may have to replace some one-time expenditures after a few years because of wear and tear.

EXPECTED YIELD PER YEAR

Check the Summer and Winter Vegetables and Herbs table for the average yield per square foot of common crops. Use the Expected Yield Per Year table in the HPAP template to estimate your yields. This will help you plan sufficient harvests for self-reliance or a homestead business.

ESTIMATED TIME

The following checklist of seasonal chores lays out the amount of time and energy a garden requires. Use it as a general guideline to fill in your weekly goals and plan your homestead calendar. Your list of chores will depend on what you raise. Customize this checklist to accommodate your needs and growing conditions.

SEASONAL CHORES FOR VEGETABLE AND HERB GARDENS

Spring

- ↗ CLEAN TOOLS: 30+ MINUTES
- ↗ TEST SOIL PH: 15+ MINUTES
- ↗ ADD COMPOST AND PREPARE GARDEN BEDS: 2+ HOURS PER BED
- ↗ START SEEDLINGS INDOORS; CHECK DAILY: 1+ HOUR
- ↗ START CROPS IN COLD FRAMES; CHECK DAILY: 1+ HOUR
- ↗ PLANT COOL-SEASON CROPS: 1+ HOUR
- ↗ HARDEN OFF TRANSPLANTS: SEVERAL DAYS
- ↗ WATER BEDS AND SEEDLINGS: 15+ MINUTES EVERY DAY
- ↗ PLANT PERENNIALS: 1+ HOUR

Summer

- ↗ PLANT WARM-SEASON CROPS: 1+ HOUR
- ↗ WATER, WEED, AND FERTILIZE CROPS: 1+ HOUR
- ↗ HARVEST, USE, AND PRESERVE PRODUCE: 1+ HOUR
- ↗ TREAT DISEASE AND PESTS: 15+ MINUTES
- ↗ PLANT COOL-SEASON CROPS FOR FALL: 1+ HOUR

Fall

- ↗ PLANT COOL-SEASON CROPS IN COLD FRAMES: 1+ HOUR
- ↗ HARVEST, USE, AND PRESERVE PRODUCE: 1+ HOUR
- ↗ GARDEN CLEANUP: 1+ HOUR
- ↗ SAVE HEIRLOOM SEEDS: 15+ MINUTES
- ↗ PLANT PERENNIALS: 1+ HOUR

Winter

- ↗ GROW MICROGREENS AND SPROUTS: 5 MINUTES DAILY
- ↗ GROW HARDY GREENS IN COLD FRAMES: 1+ HOUR
- ↗ PLAN GARDEN AND ORDER SUPPLIES: 2+ HOURS
- ↗ TEST GERMINATION RATES: 5+ MINUTES
- ↗ START COOL-SEASON CROPS INDOORS: 1+ HOUR

POTENTIAL PROBLEMS AND COST OF SOLUTIONS

Research common pests and diseases as well as prevention and treatment options. Your local Cooperative Extension Service office may offer free publications on the subject. Typical issues include aphid infestations, drought, powdery mildew, and plants destroyed by rabbits, deer, and woodchucks.

IS YOUR VEGETABLE AND HERB GARDEN PROJECT FEASIBLE THIS YEAR?

After reading through the chores checklist and the cost estimates, do you have the budget and time necessary to raise vegetables and herbs? If not, can you modify your plans to include a smaller garden? Use the HPAP template to record your ideas.

CREATING GOALS FOR VEGETABLE AND HERBS ACTION PLAN

If raising vegetables and herbs is feasible this year, it's time to set your goals. Use the breakdown tables of weekly, monthly, and first-year goals in the HPAP template to guide your plans for the first season of your vegetable and herb garden. Start with your general goals, planning tasks in manageable chunks according to the best time frame for planting, harvesting, and preserving crops. Use a weekly breakdown for busy seasons, such as spring, and a monthly breakdown for winter, when you have fewer tasks to complete.

Let's take a look at a sample weekly goal to plant a raised bed for vegetables, herbs, and pollinator-friendly flowers. This sample weekly goal assumes tools are already available. Use it to guide your step-by-step action plan.

Sample Weekly Goal: Install one 4-by-8-foot raised bed, fill with soil, and plant with vegetables, herbs, and flowers by the end of the first week of April.

DAY	TASKS TO COMPLETE	MATERIALS NEEDED	TIME	COST
MONDAY DATE: APRIL 1	⟋ Purchase materials and level the bed location	⟋ Raised bed kit (purchase) ⟋ Shovel ⟋ Level	5 to 7 hours	$80 + $5.80 tax
TUESDAY DATE: APRIL 2	⟋ Assemble the raised bed and fill it with soil	⟋ 1.2 cubic yards of topsoil (2 cubic yards delivered for $80)	4 to 6 hours	$80
WEDNESDAY DATE: APRIL 3	⟋ Rake the soil, measure out planting grids, install fencing for peas	⟋ Twine ⟋ Fencing and 4 posts ⟋ Use scrap materials	2 to 4 hours	$0
THURSDAY DATE: APRIL 4	⟋ Plant peas, lettuce, kale, collards, cabbage, broccoli, beets, turnips, green onions, cilantro, and dill seeds in bed; water them	⟋ Packets of seeds ($2+ each) ⟋ Trowel ⟋ Gloves	3 to 4 hours	$28
FRIDAY TO SUNDAY DATES: APRIL 5 TO 7	⟋ Water the plants in the raised bed		15 minutes per day for 3 days	$0
		Totals:	15 to 22 hours	$193.80

PREPARE YOUR HOMESTEAD FOR VEGETABLES AND HERBS

Many factors affect food production in your garden. You will get the best results if you improve soil health, reduce weeds and pests, water properly, and plan to increase yields. In this section, we dig into planning, crop rotation, planting in succession, building a raised bed, and choosing the best crops.

MAP OUT YOUR GARDEN

Planning your garden can be both fun and frustrating. If you're starting a new garden, choose a spot close to the house and a garden hose for easy maintenance. Make sure your garden gets at least eight hours of sunlight each day and has well-drained soil. Keep the size manageable at first—you can always expand in the future. Measure the location and outline it with rocks or twine so you can visualize the space.

You can make planting day easier by creating a detailed garden map. Sketch out your garden on graph paper and label each bed. Note where plants grew last year, if applicable. Play around with your layout until you are happy with the arrangement (it's so much easier to change plans on paper!).

Follow these steps to create your map:

1. Mark the direction of north and note the scale.

2. Draw an outline of each bed.

3. Note the locations of cold frames, fencing, trellis, and water.

4. Draw each crop on the map, leaving space for the mature size of the plant, and label it.

5. Rotate crop families according to the Crop Rotation chart below.

When you are satisfied with your layout, add it to your action plan and refer to it when you are ready to begin planting.

CROP ROTATION

Once you know the size and shape of your vegetable and herb beds, you're ready to determine the crop layout. You can also plan crop rotation, which makes better use of soil nutrients and reduces disease.

Make a list of all the plants you intend to grow. Group them according to soil and nutrient needs. The basic plant groups are LEGUMES, LEAFY GREENS, FLOWERING AND FRUITING PLANTS, and ROOT CROPS.

Use the following Crop Rotation chart to map your planting beds.

CROP ROTATION

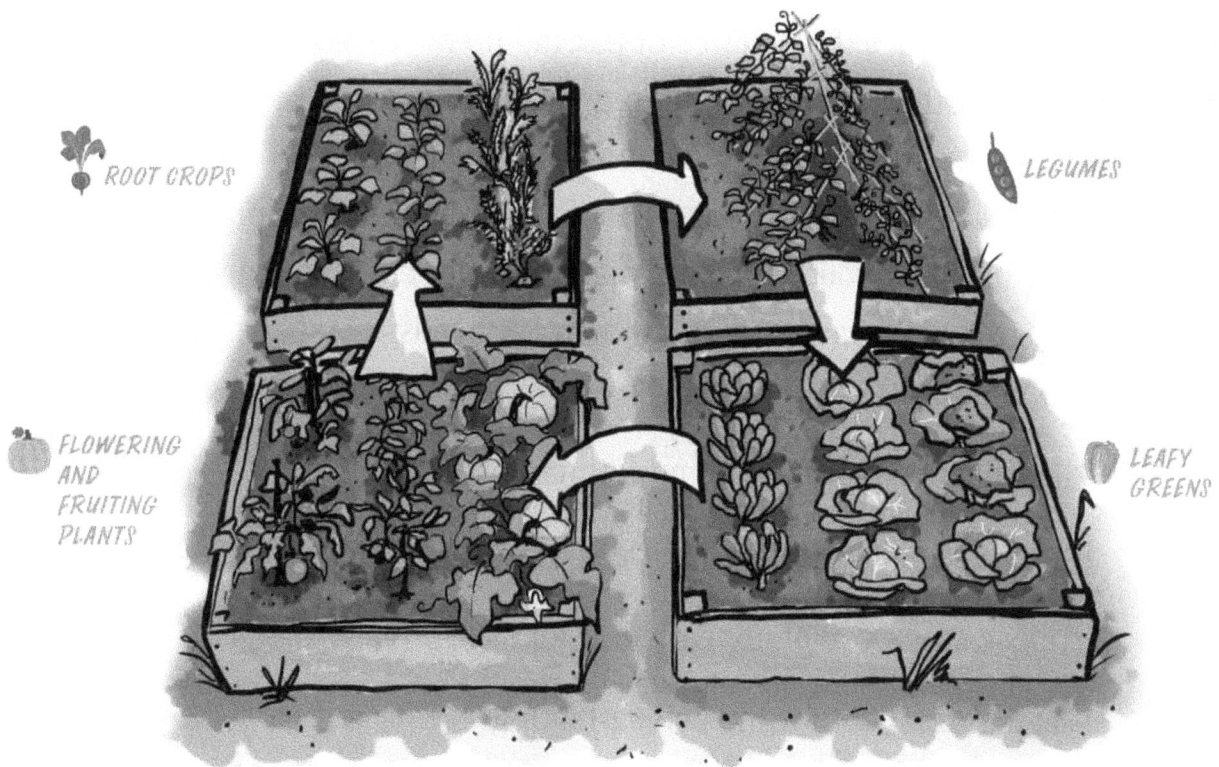

ROOT CROPS   LEGUMES

FLOWERING AND FRUITING PLANTS   LEAFY GREENS

ROOT CROPS (LOW NITROGEN NEEDS)
Beets, carrots, garlic, onions, parsnip, potatoes, radishes, rutabaga, turnips

LEGUMES (FIX NITROGEN IN THE SOIL)
Alfalfa, beans, clover, cow peas, green peas

LEAFY GREENS (HIGH NITROGEN NEEDS)

Basil, brussels sprouts, cabbage, collards, kale, lettuce, mustard greens, spinach

FLOWERING AND FRUITING PLANTS (MODERATE NITROGEN NEEDS)

Cucumbers, eggplant, peppers, pumpkins, sunflowers, tomatoes, winter squash, zucchini

Companion Planting

Companion planting is the practice of pairing up plants that provide mutual benefits to each other, such as providing shade, improving soil conditions, and keeping pests at bay. Let's look at a sample plan for a 4-by-10-foot bed using a modified three sisters (corn, beans, and pumpkins) companion planting. In this plan, you'll mix and match similar crops in a grid layout.

Customize the plants and layout to fit your needs and space. Here are some considerations:

- Plan for spring, summer, and fall harvests.

- Raise crops adapted to your area.

- Choose crops you enjoy eating or that sell well.

- Use a grid layout for intensive planting.

- Use raised beds if your soil is poor or stays wet.

- Save space by growing vines up a trellis.

- Plant space-saving varieties, such as bush pumpkins, squash, and cucumbers.

- Mulch plants with compost for increased yields.

- ◆ Plant tall crops on the north side of the bed to prevent shading sun-loving crops.

- ◆ Space plants close enough to shade out weeds without crowding each other.

SAMPLE INTENSIVE-PLANTING GARDEN MAP

1 square = 1 foot × 1 foot

4 FEET

10 FEET

PUMPKIN

SQUASH

MELON

ZUCCHINI

CUCUMBER

TOMATO

BELL
PEPPER

EGGPLANT

POLE
BEANS

SNAP PEA

CORN

SUNFLOWER

Succession Planting

Lots of gardeners plant all their crops at once. This is certainly easier to plan, but it isn't as productive as succession planting. With this technique, you sow seeds at intervals to spread out harvests.

For example, you might sow cold-hardy kale seeds in early spring, grow lettuce in midspring, and follow up with Swiss chard. This strategy provides you with greens from early spring through fall.

SAMPLE SUCCESSION PLANTING SCHEDULE

WEEKS 1, 2, 3	WEEKS 4, 5, 6	WEEKS 7, 8, 9
Plant 2 sq. ft. of kale each week (early spring)	Plant 2 sq. ft. of lettuce each week (midspring)	Plant 2 sq. ft. of Swiss chard each week (late spring)

You may also replace cool-season crops with heat-lovers after the danger of frost has passed, then start a third planting of cool-season crops in late summer for a fall harvest. Check out the table below for potential plantings.

SAMPLE SUCCESSION PLANTING STRATEGY

LATE WINTER AND EARLY SPRING	LATE SPRING AND EARLY SUMMER	LATE SUMMER AND FALL
Cool-Season Crops:	**Warm-Season Crops:**	**Cool-Season Crops:**
Beets, broccoli, cabbage, carrots, cauliflower, cilantro, dill, green onions, kale, lettuce, parsnips, peas, radishes, rutabaga, spinach, turnips	Basil, beans, corn, cucumbers, eggplant, peppers, potatoes, pumpkins, savory, squash, sunflowers, tomatoes	Beets, broccoli, cabbage, carrots, cauliflower, cilantro, dill, green onions, kale, lettuce, peas, radishes, spinach, turnips

Succession planting can get confusing, so create a plan, like the following sample, that specifies which crops to plant and when. Note the planting dates in your homestead calendar. As you work on your garden map, include crops for fresh eating and for preserving for winter.

Notes: Plant the bed with spring greens in April. Replace the greens with a main crop of a variety of tomatoes after the last frost date. Interplant with green onions and beets in fall.

- Spring lettuce: Plant April 1 to April 15.

- Spring spinach: Plant April 1 to April 15.

- Tomatoes: Start a variety of tomatoes indoors on April 1 and transplant to garden May 21 to May 28.

- Fall beets: Sow on September 1.

- Fall green onions: Sow on September 1.

SAMPLE SUCCESSION PLANTING PLAN (4-BY-10-FOOT BED)

1 square = 2 feet × 2 feet

Spring: Lettuce **Main Crop: Cherry Tomato** Fall: Beets	Spring: Lettuce **Main Crop: Rutgers Tomato** Fall: Beets	Spring: Lettuce **Main Crop: Roma Tomato** Fall: Beets	Spring: Lettuce **Main Crop: Roma Tomato** Fall: Beets	Spring: Lettuce **Main Crop: Roma Tomato** Fall: Beets
Fall: Green Onions **Main Crop: Yellow Pear Tomato** Spring: Spinach	Fall: Green Onions **Main Crop: Brandywine Tomato** Spring: Spinach	Fall: Green Onions **Main Crop: Roma Tomato** Spring: Spinach	Fall: Green Onions **Main Crop: Roma Tomato** Spring: Spinach	Fall: Green Onions **Main Crop: Roma Tomato** Spring: Spinach

BUILDING RAISED BEDS

A raised bed is a planting bed that sits above ground level. Raised beds have many advantages: They reduce problems with poor drainage and rocky soil, they warm up earlier in spring, and they are easier to tend, especially if you have trouble bending down. For easy access, limit the size of your beds to 3 or 4 feet wide.

You can build a raised bed with a variety of materials. Many are created with lumber and metal corner brackets. Kits are available, or you can use scrap wood and hold it in place with stakes. Avoid lumber treated with toxic chemicals.

Use soil rich in humus to fill raised beds. The square-foot method calls for a mixture of peat moss and other nonrenewable resources, but you can create your own soil for almost no money. Compost yard waste, manure, and kitchen scraps to fill the bed with rich black soil.

How to Build a Raised Bed

Here's how to build a wood-framed raised bed from lumber and stakes or special corner brackets for your garden.

1. Choose a level spot in full sun.
2. Measure and cut 4 pieces of rot-resistant lumber for the sides.
3. Use raised bed corner brackets or stakes to hold the sides in place.
4. Line the bottom of the bed with cardboard to kill the grass.
5. Fill the bed with soil to 2 to 3 inches from the top.
6. Use twine to mark off 1-foot squares to make planning and planting easy.

CHOOSE THE BEST VEGETABLES AND HERBS

As you plan out your vegetable and herb garden, take note of which plants are related. Plants in the same family are usually prone to the same diseases, so rotate families to prevent problems. For example, don't plant tomatoes in an area where you grew potatoes the year before.

Here are the basic plant families for vegetable and herb gardens.

Aster family. Includes chamomile, endive, lettuces, sunflowers, and tarragon. Common pests include leaf miners and aphids. Rotate crops to prevent powdery mildew.

Brassica family. Includes broccoli, Brussels sprouts, cabbage, cauliflower, Chinese cabbage, kale, kohlrabi, mustard greens, pak choi, radish, rutabaga, and turnip. Common pests include cabbage loopers, cutworms, and cabbage aphids. Rotate crops to prevent the fungal disease club root.

Carrot family. Includes caraway, carrot, celery, chervil, cilantro, dill, fennel, and parsnip. Common pests include carrot root maggots and root knot nematodes.

Cucurbit family. Includes cucumbers, gourds, melons, pumpkins, summer squash, and winter squash. Squash vine borer and squash bugs are common pests, and powdery mildew is a fungal disease often found on their leaves.

Goosefoot family. Includes beets, spinach, sugar beets, and Swiss chard. Common pests include leaf miners and earwigs. Rotate crops to prevent leaf spot and downy mildew.

Grass family. Includes barley, corn (including sweet corn), millet, oats, rice, rye, sorghum, and wheat. Common pests include corn earworms and cutworms. Rotate crops to reduce leaf blight and rust.

Legume family. Includes alfalfa, beans, clover, cow peas, peanuts, peas, and soybeans. Common pests include Japanese beetles and pea and bean weevils. Rotate crops to prevent fusarium wilt and tobacco mosaic virus.

Lily family. Includes asparagus, chives, garlic, leeks, onions, and shallots. The most common pest problem is the asparagus beetle. Rotate onions, shallots, leeks, and garlic to prevent diseases such as neck rot, onion smut, and pink root rot.

Mint family. Includes basil, catnip, lavender, marjoram, mint, oregano, rosemary, sage, and thyme. Pests and diseases aren't usually problematic.

Nightshade family. Includes eggplant, ground cherry, peppers, potatoes, and tomatoes. Common pests include tomato hornworm, potato beetles, and flea beetles. Rotate crops to prevent verticillium wilt and tobacco mosaic virus.

COOL-SEASON AND WARM-SEASON VEGETABLES AND HERBS

Most crops fit into two broad groups according to their temperature requirements. Cool-season plants thrive in cool weather and some even tolerate a light frost, but their warm-season counterparts suffer in temperatures below 50°F to 60°F. Here are some common crops in each of those categories.

Common Cool-Season Vegetables and Herbs

These crops grow well in spring and fall. Direct seed in the garden when you can work the soil. Start seeds indoors or in a cold frame for an early start.

- ↗ ASPARAGUS (PERENNIAL)
- ↗ BEETS
- ↗ BROCCOLI
- ↗ BRUSSELS SPROUTS (CAN HANDLE HEAT)
- ↗ CABBAGE
- ↗ CARROTS
- ↗ CAULIFLOWER
- ↗ CHINESE CABBAGE
- ↗ CHIVES (PERENNIAL)
- ↗ GREEN ONION (CAN HANDLE HEAT)
- ↗ KALE

- ↗ LEEKS
- ↗ LETTUCE
- ↗ MUSTARD GREENS
- ↗ PAK CHOI
- ↗ PARSNIP (CAN HANDLE HEAT)
- ↗ PEAS
- ↗ RADISHES
- ↗ RHUBARB (PERENNIAL)
- ↗ RUTABAGA
- ↗ SPINACH
- ↗ SWISS CHARD (CAN HANDLE HEAT)
- ↗ TURNIPS

Common Warm-Season Vegetables and Herbs

Plant summer-loving vegetables and herbs after the last frost. Start seeds for melons, cucumbers, squash, eggplants, peppers, and tomatoes under lights indoors and harden off (slowly increase time outside) before transplanting them into the garden.

- ↗ BEANS
- ↗ CORN
- ↗ CUCUMBERS
- ↗ EGGPLANT
- ↗ MELONS

↗ PEPPERS (BELL AND CHILE)

↗ POTATOES

↗ PUMPKINS

↗ SUMMER SQUASH

↗ SWEET POTATOES

↗ TOMATOES

↗ WINTER SQUASH

↗ ZUCCHINI

SUMMER AND WINTER VEGETABLES AND HERBS

COMMON CROPS	SPACING	YIELD PER SQ. FT.*	TIME TO HARVEST FROM SEED	BEST GROWING CONDITIONS
ASPARAGUS	**Plants:** 18 in. **Rows:** 4 to 5 ft.	0.3 lb.	3 years (from crowns)	This perennial crop takes up to 3 years to produce a harvest; plant all male crowns in full sun with well-drained soil
BEANS, GREEN	**Plants:** 2 in. **Rows:** 24 to 30 in.	0.2 lb.	50 to 60 days	Full sun Depth to plant seeds: ½ in. Time to germinate: 8 to 10 days Germination temperatures: 60°F to 75°F Direct seed: After all danger of frost
BEETS	**Plants:** 2 in. **Rows:** 24 to 30 in.	0.8 lb.	50 to 60 days	Full sun to part shade Depth to plant seeds: ½ in. Time to germinate: 10 to 12 days Germination temperatures: 45°F to 85°F Direct seed: 2 to 3 weeks before last frost
				Full sun to light shade

BROCCOLI	**Plants:** 16 to 24 in. **Rows:** 24 to 30 in.	0.36 lb.	55 to 65 days	Depth to plant seeds: ½ in. Time to germinate: 4 to 7 days Germination temperatures: 45°F to 85°F Start indoors: 9 to 11 weeks before last frost; transplant into spring garden 4 weeks before last frost
BRUSSELS SPROUTS	**Plants:** 12 to 18 in. **Rows:** 24 to 36 in.	0.2 lb.	90 to 100+ days	Full sun Depth to plant seeds: ¾ in. Time to germinate: 7 to 10 days Germination temperatures: 60°F to 65°F Start indoors: 10 to 12 weeks before last frost; transplant to spring garden 3 to 4 weeks before last frost
CABBAGE	**Plants:** 12 to 18 in. **Rows:** 24 to 30 in.	0.8 lb.	80 to 180 days	Full sun Depth to plant seeds: ½ in. Time to germinate: 3 to 4 days Germination temperatures: 65°F to 70°F Start indoors: 4 to 6 weeks before last frost; transplant to the garden after the last frost is expected
CANTALOUPE	**Plants:** 2 to 3 ft. **Rows:** 6 ft.	0.5 lb.	80 to 100 days	Full sun Depth to plant seeds: 1 in. Time to germinate: 5 to 10 days Germination temperatures: 60°F to 95°F Start indoors: 3 to 4 weeks before last frost; transplant after all danger of frost
CARROTS	**Plants:** 1 to 4 in. **Rows:** 24 in.	1 lb.	60 to 80 days	Full sun Depth to plant seeds: ¼ to ½ in. Time to germinate: 2 to 3 weeks Germination temperature: 45°F Direct seed: 2 to 3 weeks before last frost
CAULIFLOWER	**Plants:** 15 to 24 in. **Rows:** 24 to 30 in.	0.5 lb.	85 to 115 days	Full sun to partial shade Depth to plant seeds: ½ in. Time to germinate: 8 to 10 days Germination temperatures: 45°F to 80°F Start indoors: 8 to 11 weeks before last frost; transplant to spring garden 3 to 4 weeks before last frost

CUCUMBERS	**Plants:** 12 in. **Rows:** 3 to 4 ft.	0.33 lb.	50 to 70 days	Full sun Depth to plant seeds: 1 in. Time to germinate: 3 to 10 days Germination temperatures: 65°F to 90°F Start indoors: 3 to 6 weeks before last frost; transplant to spring garden after all danger of frost
EGGPLANT	**Plants:** 24 in. **Rows:** 30 to 36 in.	0.45 lb.	100 to 160 days	Full sun Depth to plant seeds: ¼ in. Time to germinate: 7 to 14 days Germination temperatures: 80°F to 90°F Start indoors: 8 to 10 weeks before last frost; transplant to spring garden after all danger of frost
GARLIC	**Plants:** 4 to 6 in. **Rows:** 24 in.	0.4 lb.	1 year (from cloves)	Full sun Plant cloves in fall 1 in. deep for harvest the following fall
LETTUCE, LEAF	**Plants:** 4 to 8 in. **Rows:** 24 in.	0.44 lb.	45 to 65 days	Full sun to partial shade Depth to plant seeds: ¼ in. Time to germinate: 7 to 14 days Germination temperatures: 40°F to 70°F Direct seed: 2 to 3 weeks before last frost in spring; plant every 2 weeks for successive harvests
ONIONS, GREEN ONIONS, STORAGE	**Plants:** 2 in. **Rows:** 24 in. **Plants:** 4 to 8 in. **Rows:** 24 in.	0.6 lb. 1.3 lbs.	20 to 30 days 100 to 175 days	Full sun Depth to plant seeds: ¼ to ½ in. Time to germinate: 7 to 10 days Germination temperatures: 68°F to 77°F Start indoors: 8 to 12 weeks before the last frost (storage onions); transplant after danger of frost; onion sets may be planted 2 to 4 weeks before last frost; direct seed green onions
PEAS, GARDEN	**Plants:** 4 in. **Rows:** 18 to 24 in.	0.09 lb.	50 to 70 days	Full sun to partial shade Depth to plant seeds: 1 to 1½ in. Time to germinate: 7 to 14 days Germination temperatures: 45°F to 70°F Direct seed: 4 to 6 weeks before last frost

PEPPERS, BELL PEPPERS, CHILE	**Plants:** 18 to 24 in. **Rows:** 24 to 30 in.	0.74 lb. 0.43 lb.	108 to 160 days	Full sun Depth to plant seeds: ¼ in. Time to germinate: 7 to 21 days Germination temperatures: 80°F to 90°F Start indoors: 8 to 10 weeks before last frost; transplant to spring garden after nighttime temperatures remain above 55°F
POTATOES	**Plants:** 8 to 12 in. **Rows:** 24 to 36 in.	1.0 lb.	10 to 15 weeks (from sets)	Full sun Plant potato sets 3 to 5 in. deep 2 weeks before the last frost; plant sets every 2 weeks for an extended harvest; hill growing plants with soil and straw to increase harvest
PUMPKINS	**Plants:** 3 to 4 ft. **Rows:** 12 in.	0.5 lb.	90 to 125 days	Full sun Depth to plant seeds: ½ in. Time to germinate: 7 to 10 days Germination temperatures: 70°F to 95°F Start indoors: 2 to 3 weeks before last frost; transplant after all danger of frost
SPINACH	**Plants:** 3 in. **Rows:** 18 in.	0.3 lb.	37 to 45 days	Full sun to partial shade Depth to plant seeds: ½ in. Time to germinate: 6 to 10 days Germination temperatures: 45°F to 75°F Direct seed: 4 to 6 weeks before last frost
SUMMER SQUASH AND ZUCCHINI	**Plants:** 2 to 3 ft. **Rows:** 3 to 4 ft.	0.75 lb.	45 to 55 days	Full sun Depth to plant seeds: 1 in. Time to germinate: 5 to 10 days Germination temperature: 60°F to 105°F Start indoors: 2 to 3 weeks before last frost; transplant after all danger of frost
SWEET CORN	**Plants:** 9 to 12 in. **Rows:** 2 to 3 ft.	0.3 lb.	60 to 100 days	Full sun Depth to plant seeds: 1 to 2 in. Time to germinate: 4 to 7 days Germination temperatures: 65°F to 85°F Direct seed: After all danger of frost
				Full sun to light shade

SWISS CHARD	**Plants:** 6 to 12 in. **Rows:** 18 to 24 in.	0.34 lb.	55 to 66 days	Depth to plant seeds: ½ to 1 in. Time to germinate: 5 to 7 days Germination temperatures: 40°F to 95°F Direct seed: 2 to 3 weeks before last frost
TOMATOES	**Plants:** 12 to 24 in. **Rows:** 24 to 36 in.	2.0 lbs.	80 to 110 days	Full sun Depth to plant seeds: ⅛ in. Time to germinate: 6 to 12 days Germination temperatures: 75°F to 90°F Start indoors: 6 to 8 weeks before last frost; transplant after all danger of frost
WINTER SQUASH	**Plants:** 6 to 12 in. **Rows:** 4 to 8 ft.	0.4 lb.	80 to 100 days	Full sun Depth to plant seeds: 2 to 3 in. Time to germinate: 5 to 10 days Germination temperatures: 70°F to 95°F Direct seed: After all danger of frost

*Yield per square foot estimates calculated from typical yields per acre from the USDA Vegetables 2018 Summary.

Tips for Beginners: Start out with a small garden and add to it as you learn. Water vegetables and herbs at the base of the plant in the morning to prevent fungal disease. Fertilize leafy greens with nitrogen and flowering and fruiting plants with phosphorus. Root crops don't usually need a lot of fertilizer. Harvest beans, peas, tomatoes, and other crops often to increase yields.

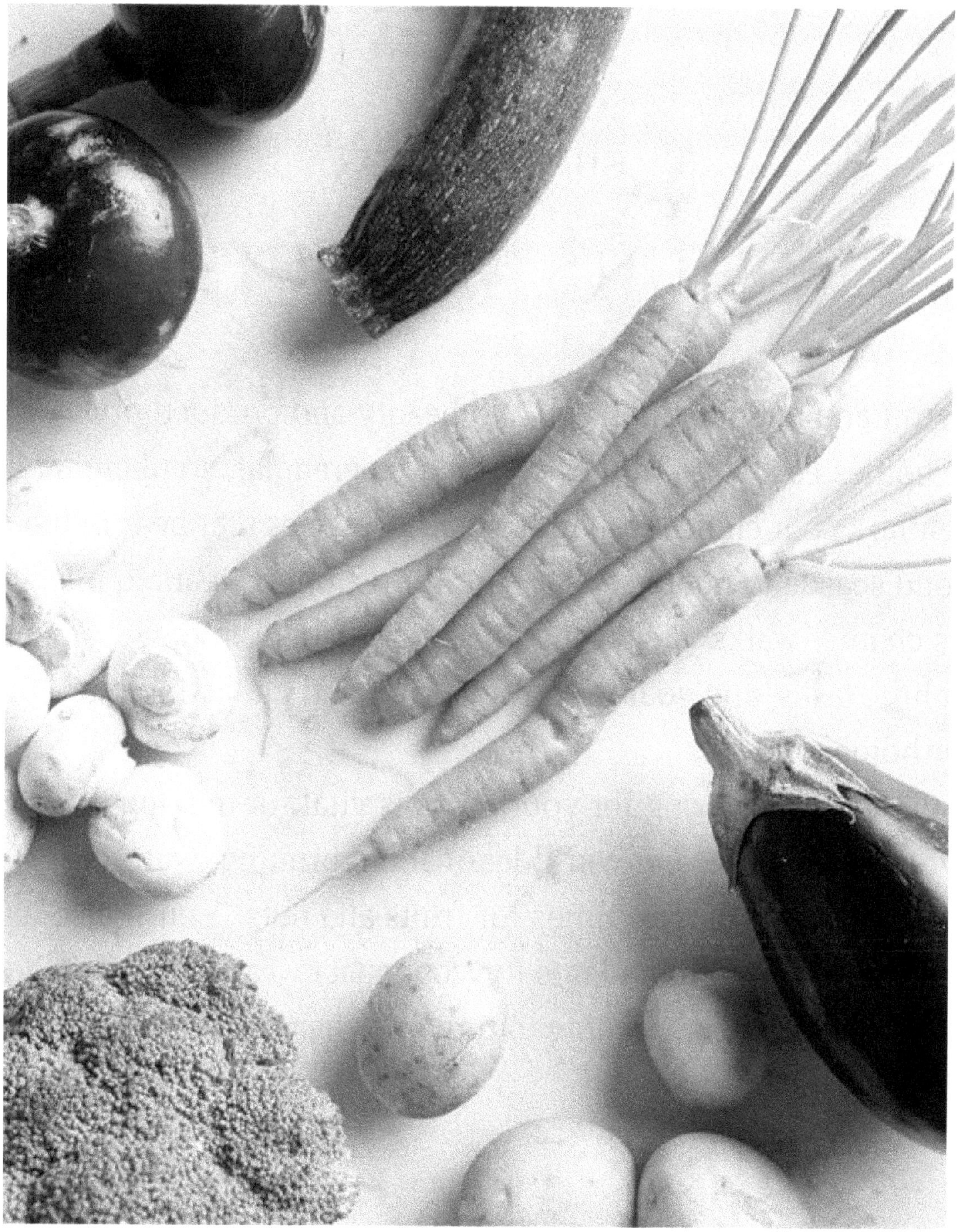

CHAPTER FOUR

FRUITS AND NUTS

Fruit and nut plantings will add beauty and productivity to your homestead. Most of these crops are perennial, so planning is essential. Your choices range from strawberries that bear in their second season to nut trees that require five or six years to mature. This chapter walks you through the steps to determine your budget, timeline, tasks, and goals to create a fruit and nut action plan for your homestead.

Planting and caring for your crops is vital for making the most of your investment, and you'll learn about pruning, watering, fertilization, and harvest times for fruits and nuts. You'll also learn how to choose the right plants for your space. Take a look at the recipes here for ideas on how to turn your harvest into delicious dishes.

CREATE A PROJECT ACTION PLAN FOR YOUR FRUITS AND NUTS

Strolling through your own home orchard is a joyful experience. Your reward for tending these plants will be fresh fruits, nuts, and preserves for winter. Use the Homestead Project Action Plan (HPAP) template to create a fruit and nut action plan for your backyard homestead.

FEASIBILITY RESEARCH

Fruit and nut plantings require research to decide whether the investment makes sense for your property. Check your homestead map to determine which fruits and nuts you can raise. If your space is limited, consider raising strawberries, blueberries, grapes, and other small fruits. Make a list of your desired crops and note their mature size and approximate cost. This will help determine the feasibility of your project and the return on your investment.

As you did with the vegetable and herb garden, it's important to gauge how much money you can potentially save by raising your own fruit and nuts rather than buying them from the local grocer.

Let's take a look at the average cost of eating fruit from the store so we can compare apples to apples. According to the USDA Economic Research Service retail price report, it costs approximately $0.72 to purchase 2 cups—the recommended daily intake for someone eating a diet of 2,000 calories—of commonly available fruits. This price assumes the following servings:

- ♦ 1 cup cored and chopped fresh apple with an average price of $0.44

- ♦ 1 cup sliced banana with an average price of $0.28

Calculate the annual cost to purchase fruit for one person by multiplying by 365: $0.72 × 365 = $262.80.

Your cost will depend on the prices in your area and the fruits you purchase. For example, strawberries are one of the easiest fruits to raise in your backyard, and the average price per serving is $0.85. If you replace a serving of bananas with strawberries from the store, your daily fruit expenditure increases to $1.29.

In your fruit and nut action plan, determine whether you have the budget and time to raise fruits and nuts successfully. Compare the expected cost of materials with your local prices, and figure out what your budget allows. Read through the seasonal chores checklist to calculate the approximate time commitment for this project.

ESTIMATED COST OF MATERIALS

There are some basic costs involved in growing fruits and nuts. Note that bare-root plants are generally less expensive than potted plants, but you must get them in the ground more quickly than potted trees. Proper implementation of organic methods can also help reduce costs.

Use the following list, your own research, and the Expected Cost of Materials table in the HPAP template to record all the materials you need for the first year. Include the cost of new plants, tree wrap, stakes, soil amendments, and fertilizer. Here are some basic costs for beginning your fruit and nut project. Typical one-time expenses are noted.*

- Fruit trees: $25 to $75*

- Nut trees: $10 to $25*

- Bramble fruits: $10 to $13*

- Blueberries: $13 to $30*

- Grapes: $15 to $50*

- Strawberries: $15 to $20 for 25 plants*

- Fertilizer: $40 to treat 6 dwarf fruit trees

- Apple maggot traps: $20+ for 6 traps*

- Horticultural oil: $16 to treat 6 dwarf fruit trees

- Sprayer: $55+*

- Tree wrap: $2.95 per 50 feet

- Pruning saw: $25+*

- Hand pruners: $12+*

Based on typical prices found online in March 2020.

*You may have to replace some one-time expenditures after a few years because of wear and tear.

Let's look at a specific example. If you chose to raise apples and strawberries, your initial investment might look like this:

- 2 dwarf apple trees: $100

- 25 strawberry plants: $20

- Fertilizer: $20

- 2 stakes for trees: $6.58

- Horticultural oil: $16

- Fruit tree sprayer: $55

Based on typical prices found online in March 2020.

In this scenario, you would expect to harvest strawberries the year after planting. Each square foot of garden space accommodates one strawberry plant that produces an average of 1.24 pounds of berries per year. The average retail cost for fresh strawberries is $2.51 per pound, so each plant could produce berries worth about $3.11, and from 25 plants, you would expect a harvest of 31 pounds of berries for a potential savings of $96.41 per year.

Each pound of berries provides three servings, so your patch could produce 93 servings of fruit in the second year. In the following table, you can see the cost of raising strawberries compared with the yields for a cost estimate of home-raised berries. When we compare this to the national average of $2.51 per pound, you see the potential to save $1.89 per pound by growing your own. Your results may vary.

COST PER YIELD ESTIMATE FOR HOME STRAWBERRY PRODUCTION

STRAWBERRY PLANTING	EXPENSES	YIELD IN LBS.
YEAR 1	$20 for 25 bare-root plants	0
YEAR 2	$10 for fertilizer	31
YEAR 3	$10 for fertilizer	31
YEAR 4	$10 for fertilizer	15.50
TOTAL	$50 investment	80.75; cost/lb. = $0.62

When estimating your cost to raise strawberries, include the price of plants ($20) and fertilizer ($30 over four years) for a total of $50. Assume an approximate yield of 80.75 pounds from years two through four. To calculate the cost per pound to raise your own berries, divide the total investment by the number of pounds produced. In this case, $50 divided by an 80.75-pound yield is $0.62 per pound of berries.

EXPECTED YIELD PER YEAR

Fill out the Expected Yield Per Year table in the HPAP template using the Choose the Best Fruits and Nuts section to help you plan. This will help you visualize the amount of food your space can produce.

ESTIMATED TIME

Another practical consideration is the amount of time it will take to plant and care for these crops. Here's a seasonal breakdown of common chores and the general time to allot for each. Use this list to estimate the amount of time necessary for your project. Note that estimates are approximate, so your time commitment will likely vary.

SEASONAL CHORES FOR PRODUCING FRUITS AND NUTS

Spring

↗ TEST SOIL PH: 15+ MINUTES

- ↗ ADD COMPOST AND PREPARE THE SOIL FOR PLANTING: 1+ HOUR
- ↗ PLANT FRUITS AND NUTS: 2+ HOURS
- ↗ FERTILIZE: 15+ MINUTES

Summer

- ↗ WATER, WEED, AND FERTILIZE AS NEEDED: 30+ MINUTES PER WEEK
- ↗ HARVEST AND PRESERVE FRUITS AND NUTS: VARIES
- ↗ CHECK FOR PESTS AND DISEASES: 15+ MINUTES PER WEEK
- ↗ FERTILIZE CITRUS FRUITS: 15 MINUTES PER TREE

Fall

- ↗ HARVEST AND PRESERVE FRUITS AND NUTS: VARIES
- ↗ RAKE LEAVES AND REMOVE FALLEN FRUITS AND NUTS: 30 MINUTES PER TREE

Winter

- ↗ WRAP TREE TRUNKS AND PROTECT FROM DEER AND RABBITS: 15 MINUTES PER TREE
- ↗ ORDER SUPPLIES: 15 MINUTES

- ↗ SPRAY TREES WITH HORTICULTURAL OIL: 15 MINUTES PER TREE
- ↗ PRUNE TREES IN LATE WINTER: 30 MINUTES PER TREE

POTENTIAL PROBLEMS AND COST OF SOLUTIONS

Research common problems you might encounter, as well as potential treatments. Your local Cooperative Extension Service office is a great place to start. Some examples include fencing off fruit trees to keep out animals, hanging apple maggot traps to reduce worm damage in fruit, and cleaning your pruning equipment to prevent the spread of fungal disease.

IS YOUR FRUIT AND NUT PROJECT FEASIBLE THIS YEAR?

After considering the seasonal chores and cost estimates, do you believe you have the time and budget to raise fruits and nuts successfully? If not, could you do a small planting the first season and add fruit and nut plants in the future? Use the HPAP template to record your ideas.

CREATING GOALS FOR FRUIT AND NUT ACTION PLAN

If raising fruit and nuts is a feasible project for your homestead, it's time to set your goals. Decide which fruits and nuts to add this year and the best time to plant them. Fill out the appropriate goal table in the HPAP template and plan the materials and time needed for planting and maintaining your project. This will help you prepare in advance and place orders in a timely manner.

Use the breakdown tables of weekly, monthly, and first-year goals in the HPAP template to guide your plans for your fruit and nut plantings. Start with your general goals, planning tasks in manageable chunks according to the best time frame for planting, harvesting, and preserving crops.

Let's look at a weekly goal sample that shows one scenario for adding fruits to the homestead, including the time and cost estimates for this project. This plan assumes tools are already available. Use this as an example to guide your step-by-step action plan.

Sample Weekly Goal: Prepare the soil, purchase, plant, and care for 2 apple trees, 2 grapevines, 6 raspberry plants, and 50 strawberry plants. Mulch with straw and water as needed.

DAY	TASKS TO COMPLETE	MATERIALS NEEDED	TIME	COST
MONDAY DATE: APRIL 1	↗ Purchase 2 potted apple trees ↗ Plant them in the new orchard area ↗ Stake the trees ↗ Water the trees	↗ Trees ↗ Shovel ↗ 2 stakes ($3.29 ea.) ↗ Twine and hose	1 hour	$100 $6.58
TUESDAY DATE: APRIL 2	↗ Purchase mulch for the apple trees ↗ Plant 50 strawberry plants ↗ Water the strawberry bed	↗ Mulch ↗ Berry plants ↗ Compost ↗ Trowel	3 hours 15 minutes	$3 $40
WEDNESDAY DATE: APRIL 3	↗ Mulch the strawberry bed with straw ↗ Water the plants, if needed	↗ 2 bales of straw	1 hour	$10
THURSDAY DATE: APRIL 4	↗ Prepare a bed for 6 raspberry plants	↗ Shovel ↗ Compost	3 to 4 hours	$0

FRIDAY DATE: APRIL 5	↗ Plant the raspberries, mulch them with straw, and water them ↗ Check plantings and water them	↗ Raspberry plants ↗ 1 bale of straw	1 hour	$66 $5
SATURDAY DATE: APRIL 6	↗ Install 2 trellises for grapevines	↗ 2 trellises ↗ 4 posts	1 hour	$30 $13.16
SUNDAY DATE: APRIL 7	↗ Plant 2 grapevines, mulch them, and water them ↗ Water fruit plantings	↗ 2 grapevines ↗ Shovel (have) ↗ 1 bale of straw	1 hour	$60 $5
		Totals:	11+ hours	$338.74

PREPARE YOUR HOMESTEAD FOR FRUITS AND NUTS

When planning for fruit and nut production in your backyard homestead, it's important to be familiar with your crops, including the length of time to harvest, approximate costs, and necessary maintenance. For example, most strawberry varieties begin producing fruit in their second year, and the largest harvests are picked in the third year. After that, your strawberry production will taper off. To continue harvesting reliably, plant a new patch every two or three years. On the other hand, an orchard requires a sizable investment, and dwarf fruit trees won't bear fruit until at least the third year. Plant strawberries for quick harvests and fruit trees for later.

MAP OUT YOUR FRUITS AND NUTS

Make a list of everything you want to plant on your homestead from the Fruits and Nuts for Backyard Homesteads table and jot down the mature size of each crop. Consult the original homestead map you created to look for potential locations for your fruits and nuts. If space is limited, consider dwarf varieties or look for creative ways to include plantings in your landscape.

Consider using hazelnuts as a windbreak, or plant almond trees for shade. You can plant dwarf fruit trees north of the vegetable garden or train them as espaliered trees (a pruning technique) along a wall. Grow grapes on a trellis or fence. Plant strawberries in a raised bed or container garden. Adjust the location of these plantings on your homestead map until you're satisfied with the layout. Take extra care when mapping—once trees and shrubs are established, you can't move them! Choose a sunny, well-drained spot and leave room for them to grow.

Follow these steps to create your map:

1. Mark the direction of north and note the scale.

2. Note the mature size of each crop you plan to grow.

3. Draw each crop on the map and label them.

4. Leave enough space to accommodate the mature size of each plant.

5. Move plants around on the map until you are happy with their location.

6. Plan for additional crops you may want to grow in the future.

When you are satisfied with your layout, add it to your action plan and refer to it when you are ready to begin planting.

CHOOSE THE BEST FERTILIZER

Fruits and nuts may need fertilization to produce ample harvests. It's a good idea to test your soil every three or four years to determine whether there are any nutrient deficiencies. Sprinkle composted yard waste and well-rotted manure around trees and small fruits to reduce fertilizer needs. There are also many organic and conventional fertilizers available at garden centers and online.

Before planting your fruits and nuts, test the soil pH. Amend it as needed to meet the following guidelines:

♦ Most fruit and nut trees and grapes prefer 6.0 to 6.5.

♦ Raspberries and blackberries prefer 5.5 to 6.5.

♦ Blueberries prefer 4.5 to 5.5.

♦ Strawberries prefer 5.8 to 6.2.

After planting your fruits and nuts, you can sprinkle a small amount of compost 2 to 3 feet from the base. Choose a balanced fertilizer with equal amounts of nitrogen (N), phosphorus (P), and potassium (K) for plants that are not growing or fruiting well. Consult the following table for the recommended rates, and follow the product's instructions for the best results.

FERTILIZER SCHEDULE FOR COMMON FRUITS AND NUTS*

CROP	PLANTING	SPRING	SUMMER	FALL	WINTER
APPLES, APRICOTS, CHERRIES, PEACHES, PEARS, PLUMS	Not necessary	1 cup of 12-12-12 per 1 in. of tree trunk diameter, up to 8 cups, in early spring	Do not apply in summer	Do not apply in fall	Do not apply in winter
NUT TREES	Not necessary	4 lbs. of compost or 4 oz. of 16-16-16 per tree applied in a ring about 2 to 3 ft. from base	Do not apply in summer	Do not apply in fall	Do not apply in winter

BRAMBLE FRUITS	Not necessary	1 to 2 lbs. of 12-12-12 per 100 ft. row until third year, then increase to 5 lbs. per 100 ft. row; fertilize before new growth	Do not apply in summer	Do not apply in fall	Do not apply in winter
BLUEBERRIES	30 days after planting, apply 1 oz. of ammonium sulfate per plant in a ring 1 to 2 ft. from base	Increase ammonium sulfate by ½ oz. per plant each year until the fifth year, then continue at that rate, or side dress annually with well-rotted manure	Do not apply in summer	Do not apply in fall	Do not apply in winter
GRAPES	Not necessary	1 lb. 12-12-12 per plant applied in a ring 2 to 3 ft. from trunk before growth starts in spring	Do not apply in summer	Do not apply in fall	Do not apply in winter
STRAWBERRIES	Not necessary	Do not apply in spring	2 to 3 lbs. of 12-12-12 per 100 ft. row in early summer	Do not apply in fall	Do not apply in winter
CURRANTS AND GOOSEBERRIES	Not necessary	1 cup 10-10-10 in a ring 2 to 3 ft. from the base, or side dress with well-rotted manure per plant	Do not apply in summer	Do not apply in fall	Do not apply in winter
CITRUS FRUITS	1 tablespoon ammonium sulfate applied around drip line of tree; water well after fertilizing	1 tablespoon of ammonium sulfate applied around the drip line from February to July; water well	1 tablespoon of ammonium sulfate applied around drip line from February to July; water well	Do not apply in fall	Do not apply in winter

PLANTING AND PRUNING

Get your fruit and nut trees off to the best start by planting them properly and watering until they are established. Prune your trees each year to encourage fruit production. Avoid working on diseased trees, which can spread fungal spores to healthy trees. These basic instructions will help you care for your orchard.

How to Plant a Fruit Tree

Early spring is the best time to plant most fruit trees. Choose a sunny location with well-drained soil. Proper planting is vital for the health of your trees. Follow these 17 steps:

1. Trim any damaged roots or branches.
2. Do not allow roots to dry out.
3. Soak tree roots in a bucket of water while you prepare the planting hole.
4. Dig the hole wide enough to accommodate the roots without bending.
5. Place the tree in the hole and make sure the graft union will be about 2 inches above soil level after planting.
6. Loosen the soil in the bottom and at the sides of the hole.
7. Do not add fertilizer or compost to the planting hole.

8. Spread the roots out over the loose soil in the bottom of the hole.

9. Put some soil over the roots, then move the tree up and down to settle the soil around the roots.

10. Continue adding soil to the planting hole, tamping it down to remove air pockets.

11. Finish filling the planting hole with soil and tamp down the soil again.

12. Water the newly planted tree to completely soak the soil and remove air pockets.

13. Stake the tree with 1 or 2 posts and a wire threaded through a 10- to 12-inch section of old garden hose (the hose prevents the wire from damaging the tree bark).

14. Mulch around the tree to hold in moisture, but do not pile mulch up the trunk.

15. Water the tree once a week unless there is adequate rainfall.

16. Continue watering every week for 4 to 5 weeks after planting.

17. Water trees during dry spells for the first season and in drought conditions thereafter.

Troubleshooting Tip: If you purchase potted trees, be sure to check for roots that circle around the root ball. These circling roots can choke off the flow of water and nutrients to the rest of the tree. Trim them before planting to encourage new roots to grow outward instead. Cut from the top to the bottom of the root ball in two or three places and in an *X* across the bottom.

How to Prune a Fruit Tree

Annual pruning encourages fruiting wood to grow on your trees. Apples, pears, cherries, and plums produce fruit best on two- to three-year-old wood, but peaches and nectarines produce best on last year's growth. Training methods include open center, central leader, and modified central leader, depending on which species you're pruning. Follow these steps:

1. Clean your pruning tools with a 10 percent bleach solution.
2. Prune before new growth begins in spring.
3. Remove any diseased or damaged branches.
4. Trim off water sprouts (vigorous branches that grow vertically) larger than ¼ inch in diameter.
5. Remove suckers from the base of the tree.
6. Begin pruning large branches by undercutting 2 to 3 inches out from the branch collar on the bottom of the branch to prevent tearing bark down the trunk when the branch drops.
7. Prune flush with the outer edge of the branch collar.
8. Do not use heading cuts, where you cut off the end of a branch.
9. If two or more branches compete for space, remove the least desirable branch completely.
10. Take frequent breaks to stand back and check your work.

Troubleshooting Tips: Branches that grow horizontally produce more fruit than those that grow up at a sharp angle. Train new branches to grow horizontally by placing a spacer bar in the crotch angle, or hang a small weight on the branch.

CHOOSE THE BEST FRUITS AND NUTS

From perennial strawberry plants to woody trees and shrubs, each group of fruits and nuts has its own characteristics and needs. Familiarize yourself with the plants you want to grow, and use this guide to help you choose the best ones for your space.

TEMPERATE FRUIT TREES

Common fruit trees such as peaches, apricots, plums, apples, pears, and cherries grow best in temperate regions with cold winters. Without a certain number of chilling hours (temperatures below 45°F) each year, these trees will not set flower buds. Some fruit trees are self-pollinating, but others require another variety. Reputable plant nurseries list the best pollinators and hardiness zones in their descriptions.

Dwarf trees may be better suited to backyard homesteads. If you have a larger space, standard-size trees provide the opportunity for sales or preserving enough food for the entire year. Use the following table to choose the best fruit trees for your homestead.

FRUIT TREES: APPROXIMATE SIZES AND YIELDS

FRUIT TREE	MATURE SIZE OF TREE	YIELD
MINIATURE	Up to 6 ft. tall and wide	¼ to 1 bushel

DWARF	6 to 8 ft. tall and wide	2 to 3 bushels Cherries: 15 to 20 quarts
SEMI-DWARF	10 to 15 ft. tall and wide	5 to 10 bushels Cherries: 30 to 50 quarts
STANDARD	25 to 30 ft. tall and wide	3 to 20 bushels

CITRUS TREES

Oranges, grapefruit, lemons, and other citrus trees grow best in southern areas. Temperatures of 30°F cause winter injury to these tender plants. Northern gardeners may grow dwarf cultivars in containers and bring them indoors to a sunny window in the fall.

Backyard homesteaders with limited space should choose dwarf or miniature cultivars. Standard-size citrus trees grow around 18 to 22 feet tall, but dwarf cultivars reach 8 to 12 feet tall and wide. Some miniature options are available for container growing and can reach 6 feet in diameter. The amount of fruit a tree produces depends greatly on its age and species.

NUT TREES

Most nut trees reach a mature size too large for a small homestead. Walnuts, butternuts, and pecans can top out at more than 75 feet tall. Walnuts and butternuts also produce a toxin (juglone) that inhibits the growth of many plants, so don't plant them near your garden or other orchard trees.

The best nut trees for small homesteads are hazelnuts (filberts) and almonds, which are technically a fruit. Expect a mature size of 10 to 25 feet tall and 10 to 15 feet wide and five years or more before nut production for both species. Hazelnuts require more than one tree for pollination, as do some almond varieties. Check the nursery description for the best pollinators.

SMALL FRUITS

This broad classification includes bramble fruits, berry bushes, grapes, strawberries, and some less common fruits. Most of these plants require less space than fruit trees, which makes them a great choice for backyard homesteads.

Strawberries are popular for backyard fruit production. These perennials usually begin fruiting in their second year and produce a harvest for two or three years. Strawberries are categorized by their fruiting season: June-bearing, everbearing, or day-neutral. June-bearing varieties produce their entire crop over two to three weeks in spring. They are further divided into early, midseason, and late varieties, and you can extend the harvest by raising some of each. June-bearing plants also produce runners for starting new plantings. Everbearing varieties produce three flushes of berries—one in spring, one in summer, and one in fall. They don't grow runners, so you have to purchase new plants or start from seed. Day-neutral varieties produce smaller harvests of berries all season long and have a few runners to start new beds.

Bramble fruits such as raspberries and blackberries have thorny canes and begin fruiting in their second year. They come in two different fruiting types: primocane (produces two crops a year, in early summer and in fall) and floricane (produces one crop a year, in early summer). There are raspberry varieties with gold, red, purple, and black fruits, each of which has a unique flavor. Thornless blackberry varieties are a good option for high-traffic areas.

Blueberries need acidic soil to thrive. If your soil is neutral or alkaline, you can grow dwarf varieties in containers or prepare their bed with ammonium sulfate the year before planting. Highbush and lowbush are native to the northern states, and rabbiteye types are best for areas with hot summers.

Grapevines are a good choice for small gardens because they grow vertically. Train them to climb a trellis, an arbor, or a fence. For commercial production, prune the vines to two or four canes to increase airflow. American grape varieties are hardier in northern areas, and European varieties need warmer growing zones. Hybrid varieties combine some of the best qualities of the American and European.

Other small fruits suitable for backyard homesteads include cranberries, currants, gooseberries, hardy kiwis, and some less commonly known fruits, like highbush cranberries.

FRUITS AND NUTS FOR BACKYARD HOMESTEADS

DWARF FRUIT TREES	SIZE	ZONES	POLLINATION	ANNUAL YIELD
APPLE	6 to 8 ft.	3 to 8	Needs a pollinator	1 to 4 bushels
CHERRY, TART	5 to 8 ft.	3 to 7	Self-fruitful	15 to 20 quarts
CHERRY, SWEET	5 to 8 ft.	2 to 7	Needs a pollinator	15 to 20 quarts
PEACH AND NECTARINE	5 to 8 ft.	5 to 7	Self-fruitful	1 to 2 bushels
PEAR	5 to 8 ft.	3 to 8	Needs a pollinator	1 to 3 bushels
PLUM	5 to 8 ft.	4 to 9	European: Self-fruitful Hybrid: Needs a pollinator	1 to 3 bushels

CITRUS FRUITS	6 to 12 ft.	8 to 11	Most are self-fruitful	Varies
NUT TREES	**SIZE**	**ZONE**	**POLLINATION**	**ANNUAL YIELD**
ALMOND	15 to 30 ft.	7 to 9	Varies by cultivar	50 to 65 lbs.
HAZELNUT (FILBERT)	10 to 20 ft.	4 to 8	Needs a pollinator	Up to 25 lbs.
SMALL FRUITS	**SIZE**	**ZONE**	**POLLINATION**	**ANNUAL YIELD**
BLUEBERRY	3 to 12 ft.	3 to 7	Self-fruitful but produce better with cross-pollination	5 to 20 lbs.
GRAPE	Prune to fit your space	American: 5 to 10 European: 7 to 11	Most are self-fruitful	10 to 12 lbs.
RASPBERRY/BLACKBERRY	2 to 3 ft.	4 to 9	Self-fruitful	1 to 2 quarts
STRAWBERRY	8 to 12 in.	4 to 9	Most are self-fruitful	1½ lbs.

GRAPES

Grapes are a versatile and delectable fruit for backyard homesteads. Vines are most productive when trellised and pruned to about 6 feet wide. Plant grapes in a location with well-drained soil and at least eight hours of sun a day. Once established, they can withstand periods of drought. Check the USDA Plant Hardiness Zone and read the requirements before selecting your grapes.

The flavor and quality of your grapes depends a great deal on your growing conditions and the climate in a given year. Rainy seasons, shade, and overgrown vines encourage fungal disease and reduce sugar content.

Select the best varieties for your needs, plant them in an ideal location, and use proper pruning, watering, fertilizing, and harvesting methods to make the most of your vineyard.

You may need to use a fungicide on grapevines to prevent disease and netting to protect your ripening fruit from birds. Harvest grape bunches as the fruit ripens. Clusters don't ripen all at once, and the grapes at the end ripen last. Taste grapes as you harvest, and use or preserve them quickly to prevent spoilage.

The following table lists common grapes for different climates. Check with your local Cooperative Extension Service office for recommendations on the best varieties and information about diseases and pests in your area.

COMMON GRAPEVINE VARIETIES

VARIETY	COLOR AND CHARACTERISTICS	ZONE	BEST USES	HARVEST SEASON
CANADICE	Red seedless	5	Juice, jelly, rosé wine, table	Very early
CATAWBA	Red, cold hardy	4	Juice, jelly, rosé wine, table	Late
CONCORD	Blue/black, cold hardy	4	Juice, jelly, rosé wine, table	Late to midseason
DELICIOUS	Blue/black, muscadine for southern areas	7	Red wine, table	Early
DIXIE	White, muscadine for southern areas	7	Juice, jelly, white wine, table	Midseason
NIAGRA	White, susceptible to disease	5	Juice, jelly, white wine, table	Midseason
RELIANCE	Red seedless, susceptible to disease	5	Juice, jelly, rosé wine, table	Early to midseason

Tips for Beginners: Choose disease-resistant varieties and clean up all dropped fruit and leaves in fall to reduce insect and fungal damage. Water fruit and nut plantings during times of drought. Harvest pears and other soft fruits before peak ripeness to prevent rot.

CHAPTER FIVE

GRAINS AND LIVESTOCK FEED

Do you dream of amber waves of grain in your own backyard? Perhaps you want to grow wheat to grind your own flour for home-baked bread. Raising your own grain and livestock feed might seem like a prospect for large farms, but it is possible to include a small-scale planting on an average-size lot. You can grow and harvest alfalfa, sunflower seeds, and other crops to feed your bees, chickens, goats, and family.

As you work through this chapter, you'll create a grain and livestock feed action plan, and we'll discuss the cost of raising small grains, as well as how to choose the best crops, how to prepare the soil, how much seed to purchase, and tips for maintaining and rotating your crops. There are so many ways to use homegrown grains in the kitchen, and you can find simple, delicious recipes here.

CREATE A PROJECT ACTION PLAN FOR YOUR GRAINS AND LIVESTOCK FEED

Most Americans leave the task of raising grain and livestock feed to the farmers, but you can produce enough grain in your backyard homestead to feed your chickens, supplement your chicken feed, or make freshly baked goods for your table.

Use the Homestead Project Action Plan (HPAP) template to create an action plan for grains and feed. This section covers the specific considerations that will guide your research and assessment.

FEASIBILITY RESEARCH

The first step is calculating how much space you have and choosing the grains you'd like to grow. Grains can work well in a crop rotation plan with your vegetables. An average backyard homestead has the potential to put 800 to 1,000 square feet of growing space into grains and feed crops. Grow alfalfa to feed livestock and improve the soil for field corn the following year. Plant winter wheat in fall to harvest next summer or plant rye as a green manure or pasture for grazing.

It's helpful to know what grows well in your area, and your local Cooperative Extension Service office has a wealth of information that can help you choose the best crops, set goals, and create a workable action plan.

It's also important to have a sense of the cost of purchasing grain versus raising it yourself. Let's use the information in the Expected Cost of Materials section to do some quick math.

According to the Cost Estimate to Plant Common Grains and Livestock Feed table, wheat is planted at the average rate of 26 seeds per square foot. If you plant 1,000 square feet of wheat, you'll need 26,000 seeds. A pound of wheat has approximately 15,000 seeds, so you'll need 1.7 pounds of wheat seed. Assume a purchase of two pounds of seed for this project.

Assuming an average yield of 0.07 pounds of wheat per square foot from 1,000 square feet, expect to invest $18 for a yield of 70 pounds of wheat. Compare this with the price of a 50-pound bag of wheat berries at $22.50 or $0.45 per pound from a food co-op in 2020. That means 70 pounds of wheat berries would cost $31.50. When you subtract the $18 investment, you have saved $13.50. If you can't grow wheat, you could potentially buy wheat berries to grind your own flour.

How much bread can you make from 70 pounds of wheat berries?

♦ 1 pound of bread flour is about 3½ cups.

♦ 1 pound of wheat berries makes about 4½ cups of flour.

♦ 70 pounds of wheat makes about 315 cups of flour.

♦ 1 loaf of homemade bread requires about 3 cups of flour.

If you have 315 cups of flour and you use about 3 cups per loaf, you could make about 105 loaves, or about two loaves a week for one year. A five-pound bag of whole-wheat flour from one major retailer costs $3.76 at the time of this writing, which is approximately $0.75 per pound. That means 70 pounds of flour would cost around $52.50. When we subtract the cost to produce 70 pounds of wheat, we have a savings of $34.50.

In the grain and livestock action plan, determine whether you have the budget and time to raise grains and livestock feed successfully. Compare the expected cost of materials to your local prices, and calculate what you can afford. Read through the seasonal chores checklist to get a handle on the time commitment for this project, and be realistic about whether you can take on the necessary labor.

EXPECTED COST OF MATERIALS

Cost is an important consideration of producing these crops at home. Using this section's cost tables, your own research, and the Expected Cost of Materials table in the HPAP template, list all the materials you need to set up and maintain the project for the first year. Include seeds, soil tests, tools, and soil amendments. If organic grain production is not a priority, you may wish to add fertilizers and pesticides to your list.

    The following table lists the average number of seeds to plant per square foot for each crop, the typical cost of one pound of seed in 2020, and the average number of seeds per pound to get an idea of the cost to plant these crops (rounded to the nearest penny). The last two columns include the average yield per square foot (from USDA records or Cooperative Extension Service offices around the United States) and the final cost to produce each pound of that specific crop, in some cases less than a penny per pound.

    The following figures are based on average seeding rates, prices, and yields, so your cost analysis may differ. Keep in mind that some varieties have different seeding rates. Use this table as a starting point for calculating the basic expenditures for grain production.

COST ESTIMATE TO PLANT COMMON GRAINS AND LIVESTOCK FEED

CROP	SEEDS/SQ. FT.	COST OF SEED/LB.	SEEDS/LB.	COST TO PLANT 1 SQ. FT.	YIELD LBS./SQ. FT.	COST/LB. TO PRODUCE
ALFALFA	90.00	$11	200,000	$0	0.18	$0.03

BARLEY	30.00	$3.20	14,300	$0.01	0.10	$0.07
BUCKWHEAT	15.00	$6.15	14,800	$0.01	0.02	$0.31
CLOVER	30.00	$20	776,000	$0	0.11	$0.01
CORN	1.00	$4.40	1,300	$0	0.20	$0.02
HAY	74.00	$4.80	540,000	$0	0.10	$0.01
OAT	28.00	$11	14,500	$0.02	0.04	$0.53
SOYBEAN	3.00	$1.95	2,500	$0	0.07	$0.03
SUNFLOWER	0.50	$2.75	18,500	$0	0.03	$0
WHEAT	26.00	$1.50	15,000	$0	0.07	$0.04

Seeding rates and number of seeds per pound vary according to the variety planted. Based on typical prices found online in March 2020.

To calculate how much seed you'll need to purchase, multiply the number of square feet to plant by the average number of seeds per square foot: *number of square feet to plant × number of seeds per square foot = number of seeds needed.*

Determine how many pounds you need by dividing the total number of seeds needed by the average number of seeds per pound: *number of seeds needed / number of seeds in one pound = number of pounds to purchase.*

Next, figure out the additional supplies you need to raise your own field crops. Here is a sample seasonal breakdown of the cost estimate for raising 1,000 square feet of Hard Red Spring Wheat, an excellent choice for making bread. Your costs will vary depending on what tools you need to purchase.

- ♦ Soil test: $15

- ♦ 2 pounds of wheat seed: $3

- ♦ Compost: $0

- ♦ Rototiller rental: $50

You can raise grains organically by planting disease-resistant varieties, practicing crop rotation, and adding compost each year.

EXPECTED YIELD PER YEAR

Using the Grains for the Backyard Homestead table and your homestead map, determine how much space you will use for raising these crops. Fill out the Expected Yield Per Year table in the HPAP template to calculate your potential yields for your project. Will they fulfill your needs for the year or do you need to increase the space for your project? Which crops offer the best return for your investment?

ESTIMATED TIME

For an idea of how much time and energy raising grain requires, check out the following seasonal chores checklist. Use this list to fill out the Estimated Time section of your grain and livestock feed action plan.

The following checklist helps you estimate the amount of time you will need to spend planting, caring for, and harvesting grain and feed crops on a 1,000 square-foot plot. Your time will vary depending on the size of your planting and the methods used.

SEASONAL CHORES FOR RAISING WHEAT

Spring

- ↗ TEST SOIL PH: 15+ MINUTES
- ↗ ADD COMPOST: 1+ HOUR
- ↗ TILL THE SOIL: 2+ HOURS WITH ROTOTILLER
- ↗ BROADCAST SEED BY HAND AND RAKE: 2+ HOURS
- ↗ IRRIGATE WITH A SPRINKLER UNTIL THE SEEDLINGS ARE ESTABLISHED: 1+ HOUR PER DAY AS NEEDED

Summer

- ↗ WEED AND WATER AS NEEDED: 1+ HOUR PER WEEK (FOR 4 MONTHS)
- ↗ CHECK FOR DISEASE: 15 MINUTES PER WEEK

Fall

- ↗ CHECK FREQUENTLY FOR RIPENING: 15+ MINUTES PER WEEK
- ↗ HARVEST THE WHEAT BY HAND: 4+ HOURS
- ↗ SHOCK THE WHEAT (BUNDLE TO DRY): 4+ HOURS
- ↗ THRESH THE WHEAT BY HAND: 4+ HOURS

↗ WINNOW THE WHEAT BY HAND: 4+ HOURS

↗ REMOVE THE HULLS AND STORE: 6+ HOURS

POTENTIAL PROBLEMS AND COST OF SOLUTIONS

Research and jot down common diseases and pests and possible solutions with the help of your local Cooperative Extension Service office. Common issues are pests such as weevils, rust and other fungal infections, and weather-related issues such as drought.

IS YOUR GRAIN AND LIVESTOCK FEED PROJECT FEASIBLE THIS YEAR?

After considering the seasonal chores and cost estimates, do you believe you have the time and budget to raise grain and livestock feed successfully? If not, can you modify your plans to include a smaller grain project or plan for grains in the future? Use the HPAP template to record your ideas.

CREATING GOALS FOR GRAINS AND LIVESTOCK FEED ACTION PLAN

If raising grain and livestock feed is a feasible project for your homestead, it's time to set your goals. Make sure they align with your overall homestead goals. For example, if you want to increase your self-sufficiency, which grains and feed are most important? Perhaps you want to raise grain to supplement your chicken feed. Record your overall goals, then break them down into steps.

Use the breakdown tables of weekly, monthly, and first-year goals in the HPAP template to guide your plans for your grain and feed crops for your first season. Start with your general goals, planning tasks in manageable chunks according to the best time frame for planting, harvesting, and processing crops. Be realistic, and add your weekly and monthly goals to your homestead calendar so you know what you need to accomplish each step of the way.

Let's take a look at a weekly goal sample to plant corn to supplement your livestock feed for winter.

Weekly Goal: Till soil and plant 1,000 square feet of dent corn for livestock feed.

DAY	TASKS TO COMPLETE	MATERIALS NEEDED	TIME	COST
MONDAY DATE: MAY 21	↗ Till the soil and spread rotted manure on surface	↗ Rototiller (have) ↗ Manure (have)	6 to 10 hours	
TUESDAY DATE: MAY 22	↗ Till the manure into the bed	↗ Rototiller	4 to 6 hours	
WEDNESDAY DATE: MAY 23	↗ Rake the bed by hand ↗ Plant the corn seed with a hand planter ↗ Put the sprinkler on the bed to water it	↗ Corn seed ↗ Hand planter (have) ↗ Sprinkler (have)	4+ hours	$4
THURSDAY DATE: MAY 24	↗ Irrigate the bed with the sprinkler		15 minutes	
FRIDAY DATE: MAY 25	↗ Check the bed and irrigate if needed		15 minutes	

SATURDAY DATE: MAY 26	↗ Check for germination and irrigate		15 minutes	
SUNDAY DATE: MAY 27	↗ Check for germination and irrigate		15 minutes	
		Totals:	15 to 21 hours	$4

PREPARE YOUR HOMESTEAD FOR GRAINS AND LIVESTOCK FEED

Planting a stand of heritage grains or livestock feed is a satisfying project. Be sure to choose a sunny, well-drained spot, prepare the soil properly, and plan for weeding, watering, and checking for pests and disease. You'll also want to be ready for harvest season with the essential tools for harvesting, processing, and storing your grains. This section explores how to get more grains from your space, rotate your crops, and best care for your plantings.

ASSESS YOUR SPACE

The amount of space you need for planting grains and feed crops depends on how much you want to harvest and the crops you choose. In the previous section, we found that 70 pounds would be a good harvest on 1,000 square feet. If you have 500 square feet for grain production, you have the potential to raise about 35 pounds of wheat.

Perhaps you can spare only 250 square feet of space for raising grain and feed crops. This could provide a harvest of about 20 pounds of oats, 50 pounds of corn, or 17.5 pounds of wheat. If you want to grow the maximum harvest from that space, corn is the best cereal grain, but you could also raise sugar beets, an even more productive crop. In this same space, you have the potential to produce around 288 pounds for fattening pigs or as a supplemental feed for poultry. You can even make your own sugar by pressing the roots and cooking down the juice.

Most backyard homesteaders don't have enough space to raise all the grains and feed necessary for the year, but you can grow enough to supplement your livestock feed or enjoy some freshly baked goods created with your own artisan grains.

MAP OUT YOUR GRAINS AND LIVESTOCK FEED

Choose large, sunny areas of your property for raising grain and feed crops, then create a map to help you make the most of your homestead. The map will also help you plan how to rotate crops from season to season to improve soil fertility and reduce pest and disease problems.

Grab your homestead map and look for areas large enough to raise a stand of grain or livestock feed. Could you replace your lawn with clover for your bees or to use as livestock forage? Is it possible to rotate your garden beds and raise grain every other year? When you're ready, map out your grain and feed crops.

Follow these steps to create your map:

1. Mark the direction of north and note the scale.

2. Draw each grain or feed crop area to scale.

3. Look at the list of crops you wish to grow.

4. Note the average yield and uses for these crops.

5. Choose crops that work best for your growing conditions and needs.

6. Add the crops to the map.

7. Keep your map with your grain and livestock action plan to track future crop rotation.

Once you've created your map and developed a working plan for your grain and livestock feed production, it's time to get growing.

PREPARE THE SOIL AND BUY SEEDS

Most grains will grow and produce a decent harvest if you till a healthy dose of composted manure into the soil before planting. Till the soil well or choose a no-till method and kill the turf before planting. Do a soil test to check the pH levels and soil nutrients, and amend the soil if necessary.

Another way to increase soil fertility is to raise alfalfa, clover, soybeans, or another leguminous crop the first year. They provide nectar and pollen for honeybees and fix nitrogen to feed your crops the following year.

Small stands of grain usually require just one or two pounds of seeds. You can find a large selection of crop species and varieties for sale online, from small seed packets to 25- or 50-pound bags, or check with your local feed supply store.

CHOOSE THE BEST GRAINS AND LIVESTOCK FEED

It's important to choose the best crop for your growing conditions, prepare the soil properly, and plant seed at the optimum time. Your growing conditions will dictate which crops you can grow and how much they will yield. In southern areas with adequate rainfall, you can grow long-season crops such as sorghum. Northern growers with cold, wet soil in the spring can plant oats early for a summer harvest. Follow up with a fall planting of winter wheat or rye to harvest the following summer. If you have a short season, choose a crop that matures early, such as barley or buckwheat. Corn, wheat, alfalfa, sunflower, and sugar beets grow well in many regions.

GRAINS FOR THE BACKYARD HOMESTEAD

GRAIN	SEASON	HOW TO USE	YIELD LBS./SQ. FT.	NOTES
BARLEY	Spring planting Summer harvest	Beer, cereal grain, livestock feed	0.1	Northern crop Hull before feeding to livestock
BUCKWHEAT	Late spring to midsummer planting Summer to fall harvest	Cereal grain, livestock feed (in small amounts), source of nectar for bees	0.02	Short season allows for another crop before or after harvest Hull before using
CORN	Late spring planting Late summer to fall harvest	Cereal grain, livestock feed	0.2	Heavy feeder Needs warm weather
OATS	Spring planting Summer harvest	Cereal grain, livestock feed, straw	0.04	Hull before using Tolerates cold, wet soil
WHEAT, HARD RED WINTER	Fall planting Summer harvest	Cereal grain, livestock feed, cover crop	0.07	Higher gluten content is good for bread
WHEAT, SOFT WHITE WINTER	Fall planting Summer harvest	Cereal grain used for pastries and cakes, livestock feed	0.07	Low gluten content, soft texture, white color

Yield estimates calculated with figures from the USDA Crop Production 2018 Summary.

OTHER FIELD CROPS FOR THE BACKYARD HOMESTEAD

CROP	SEASON	HOW TO USE	YIELD LBS./SQ. FT.	NOTES
ALFALFA	Spring planting for harvest in late summer, fall, or subsequent years	Livestock feed, cover crop, soil conditioning, nectar for bees	0.18	Alfalfa fixes nitrogen in the soil for future crops; soil pH should be 6.5 to 7
BEETS, SUGAR	Spring planting Fall harvest	Sugar production, livestock feed, human consumption	1.15	Sugar beets are high in sugar and are good for finishing pork or feeding livestock through the winter
CLOVER	Spring planting for harvest in late summer, fall, or subsequent years	Livestock feed, cover crop, soil conditioning, nectar for honeybees	0.11	Clover is a great cover crop and provides forage or hay for livestock as well as a food source for bees
HAY, ORCHARD GRASS	Perennial crop	Livestock feed	0.1	Hay is typically harvested for several years before rotating to leguminous crop
SOYBEANS	Spring planting Fall harvest	Livestock feed, human consumption	0.07	Soybeans are often used in crop rotations with field corn in agricultural areas. Must be roasted before consumed by humans or livestock
SUNFLOWER, BLACK OIL	Spring planting Fall harvest	Livestock feed, human consumption, birdseed	0.03	Black oil sunflowers are easy to grow and provide fat and protein for livestock

Yield estimates calculated with figures from the USDA Crop Production 2018 Summary.

HOW TO MILL GRAINS FOR FLOUR

Grinding your own grain gives you the freshest flavor and most nutrient-dense baked goods, cereals, and pastas you'll ever eat. Whole grains store well, sometimes for years, without losing their nutritional value. Flour, on the other hand, begins to lose nutrient content and flavor as soon as you grind the grain. For the best flavor, grind your flour just before baking.

You don't necessarily need a grain mill to grind your own flour and cornmeal. A regular blender will do the job in a pinch, but you might wish to invest in a grain mill if you intend to grind grain regularly. Too much grinding can damage your blender, and a mill does a much better job of grinding to the desired consistency.

How to Grind Your Own Flour

Here's how to use a kitchen blender to grind your own flour from wheat berries:

1. Put the wheat berries in the blender.
2. Start the blender on the lowest setting.
3. Blend for 1 minute, then turn off the blender.
4. Check the consistency of the flour.
5. Blend again if the flour is too coarse.

6. Don't overgrind the flour, as the heat of the blender can damage the flavor.
7. Grind only as much flour as you can use right away.
8. Sift the flour to remove any large pieces of grain, then regrind them.

Grain mills come in manual and electric varities and as attachments for stand mixers. Here are some qualities to look for:

↗ EASY TO TAKE APART AND CLEAN

↗ MADE WITH QUALITY PARTS

↗ A GOOD REFUND OR WARRANTY POLICY

↗ A VARIETY OF SETTINGS, FROM FINE TO COARSE

Compare reviews and read the description before purchasing your new grain mill to make sure it fits your needs.

Troubleshooting Tip: Make sure your grains are properly dried before storing to prevent mold. Store your whole grains, in an airtight container to prevent damage from Indian meal moths and other pests, in a cool (under 70°F), dry, dark spot. For long-term storage, put the grains in the freezer.

Tips for Beginners: You'll need to pull weeds until your grain crops are established for best results. Increase soil fertility with leguminous crops such as field peas or soybeans, or raise a cover crop and till under for green manure. Sugar beets have low nitrogen requirements, and pigs will happily harvest the crop for you. Just fence in your beet planting and allow the pigs to root up their dinner, tilling the soil and adding manure as they go.

CHAPTER SIX

RAISING POULTRY

Keeping backyard chickens has become increasingly popular in recent years. Poultry can provide your family with farm-fresh eggs and humanely raised meat for the table. With an average-size lot, you can keep a small laying flock and raise broilers for meat each year. Ducks also provide delicious eggs, and meat turkeys make wonderful holiday meals. Poultry are well adapted to life on a small homestead, and the cost is relatively low compared with most livestock.

Before starting a flock, check local regulations and read through this chapter to determine the cost of keeping poultry, the seasonal needs of your flock, the time commitment, and which species and breeds will best suit your needs.

This chapter helps you work through everything you need to consider in order to determine the feasibility of keeping chickens, ducks, turkeys, and other poultry on your backyard homestead. We've also included recipes featuring your eggs and poultry here.

CREATE A PROJECT ACTION PLAN FOR YOUR POULTRY

What better way to get back to basics than gathering eggs? Chickens are often the first livestock that backyard homesteaders consider raising, and for good reason. Keeping a small flock of chickens is a wonderful way to increase your self-reliance and maybe even earn a side income. If you want to save money, get eggs and meat from your homestead, or keep poultry for the small farm experience, start by researching the costs involved, time required, and best practices.

Use the Homestead Project Action Plan (HPAP) template to create a poultry action plan for your backyard homestead. Here are the specific considerations that will guide your research and assessment.

FEASIBILITY RESEARCH

To assess your poultry project, you'll need to determine the up-front investment and time required to care for your birds. There are many variables to consider, including how many you can accommodate in your space, whether you're raising them for eggs or meat, and the cost of feed and materials such as feeders and coops.

Start by determining whether it's cheaper for you to raise poultry yourself or purchase eggs and meat from a local grocer.

Let's consider how much it might cost to raise farm-fresh eggs compared with purchasing them. The most productive hens can lay between 280 and 300 eggs per year. That means 25 layers could produce 7,000 to 7,500 eggs annually, or 583 to 625 dozen eggs.

By combining the cost to raise 25 chicks to point of lay and the cost to feed 25 hens per year (calculated later in this chapter), we can estimate the investment for 25 chickens for the first 17.5 months, about the time of their first molt, when they take a break from production. For that time frame, it would cost $840.10 to $2,218.35 for their basic care. That means if they lay 625 dozen eggs, it would cost between $1.34 and $3.55 to produce each dozen.

According to the egg industry, the average cost of a dozen eggs in the United States at the time of this writing was $2.73, which means a potential savings of $1.39, or a loss of $0.82, per dozen eggs. If your family uses two dozen eggs per week, you could save $139.36 each year if you keep expenses to a minimum. But based on these estimates, you could also potentially lose $85.28 when raising heavy breed hens on expensive feed.

How much you save depends on many factors, from the cost of chicks, feed, bedding, and a coop to the electricity required to power a brooder and keep a light on in winter. Here are some creative ways to reduce expenses:

♦ Raise grains and produce to supplement the chicken feed.

♦ Raise the chickens on pasture or feed them with weeds and kitchen scraps.

♦ Raise light breeds or hybrids for best production and feed consumption.

♦ Sell extra eggs, chicks, hatching eggs, or point of lay pullets.

♦ Process old laying hens for stewing to reduce expenses.

♦ Keep only productive hens and a rooster (if you want fertile eggs).

Whether you raise poultry for meat or eggs, you'll need to plan. It takes four and a half to five and a half months to raise pullets (young female chickens) to point of lay, or the age when they should begin to produce eggs. Broiler chickens and meat ducks take about eight weeks, and broad-breasted turkeys need about five months to reach butcher weight.

In your poultry action plan, determine whether you have the budget and time to raise poultry successfully. Compare the expected cost of materials with your local prices or the current cost of products online, and figure out what your budget allows. Read through the seasonal chores checklist to determine the approximate time commitment for this project.

EXPECTED COST OF MATERIALS

As you research the feasibility of this homestead project, be sure to take into account the cost of purchasing and caring for poultry. Using this section's cost tables, your own research, and the Expected Cost of Materials table in the HPAP template, list all the materials you need to set up and maintain the project for the first year. Include the cost of the first chicks, feed, bedding, coop, and other equipment.

Many people choose to start their poultry journey by investing in chicks. This list outlines the average cost of day-old birds.

Cost of Day-Old Poultry:

- Chickens: $1.50 to $6

- Ducks: $4 to $10

- Geese: $12 to $30

- Turkeys: $5.25 to $13

- Guinea fowl: $4 to $8

- Quail: $2.50 to $4

Keep in mind that until they are ready to go into their coop, chicks need a protected box or brooder, which should be part of your cost considerations. A large plastic storage tub may suffice for a few chicks, or you can build a brooder box to house them. The brooder should be big enough to provide the chicks ample room to move about and find the most comfortable temperature. To ensure a healthy start, you'll also need to provide chick starter feed, chick grit, clean water, and probiotics.

Cost of 50 Pounds of Poultry Feed (cost per pound):

- Chicken layer: $12 to $32 ($0.24 to $0.64)

- Chick starter: $13.50 to $30 ($0.27 to $0.60)

- Chicken meat producer: $14 to $28 ($0.28 to $0.56)

- Game bird: $16 to $27 ($0.32 to $0.54)

- Oyster shell (layer supplement): $12 to $33 ($0.24 to $0.66)

Use the following table to estimate feed costs and get a sense of how much it costs to feed one adult laying hen per day.

COST OF FEEDING LAYING HENS IN FULL PRODUCTION IN 2020

LAYER TYPE	LBS. OF FEED PER DAY PER HEN	FEED COST PER DAY PER HEN
LIGHT BREED	0.24	$0.06 to $0.15

HEAVY BREED	0.30	$0.07 to $0.19

If you plan to feed a flock of 25 laying hens, expect to spend $1.50 to $4.75 each day, or $547.50 to $1,733.75 per year, assuming a price of $12.50 per 50 pounds of layer feed.

Unless you purchase point-of-lay pullets, you'll also need to calculate the cost of raising chicks until they begin producing. The following table breaks down the cost of raising 25 chicks, assuming average prices for chicks and basic supplies.

COST TO RAISE 25 CHICKS TO POINT OF LAY AT 22 WEEKS

ITEM	LIGHT BREED	HEAVY BREED
DAY-OLD PULLET ($4)	$100	$100
FEED (16 LBS. FOR LIGHT BREED, 20 LBS. FOR HEAVY BREED)	$108 to $240	$135 to $300
BEDDING ($4.59 × 4)	$18.36	$18.36
FEED AND WATER CONTAINERS	$8	$8
HEAT LAMP AND BULB	$11	$11
ELECTRICITY	$30.24	$30.24
ELECTROLYTES AND PROBIOTICS	$8	$8
GRIT	$9	$9
TOTAL COST	$292.60 to $424.60	$319.60 to $484.60
COST TO RAISE PER CHICK	$11.70 to $16.98	$12.78 to $19.38

You may use some items for years, such as the feed dishes and heat lamp, but there are other expenditures to consider.

Other Poultry Costs:

- Chicken coop (homemade or purchased): $50 to $1,000

- Fencing (50-foot roll): $29 to $130

- Fence posts: $3.29 to $5

- Waterer: $6 to $35

- Feeder: $6 to $30

- Water heater for winter: $30 to $60

- Nest box: $13 to $20

- Pine shavings (8.25 cubic feet): $4.59 to $9

- Straw: $5+ per bale

- Vitamins and probiotics: $12 to $20 per 2 pounds

- Egg cartons: $0.39 to $0.54 each

- Grit: $8 to $9 per 25 pounds

- Oyster shell: $12 to $15 per 50 pounds

There are various ways to reduce the cost of supplies. With some basic tools, you can build your coop, roosting bars, and nest boxes from scrap lumber or retrofit a garden shed or small barn.

Next, let's look at the cost of consumable items needed to care for a flock of 25 hens for one year. Your hens will consume between 2,190 and 2,737.5 pounds of feed for a cost of $547.50 to $1,733.75. They'll also need about 50 pounds of grit at $16 to $18, and up to 100 pounds of oyster shell for $24 to $30. You'll need to provide 12 to 24 bales of straw bedding at $60 to $120. The basic cost total is $647.50 to $1,901.75 to keep 25 laying hens each year.

To calculate the cost for each hen, divide the total costs by 25 for an estimate of $25.90 to $76.07 per bird. You can use this calculation to estimate consumables costs for any size flock.

To calculate one-time expenses in your cost analysis, total them up and divide the cost by the life expectancy of the materials. Add these expenses to your egg and meat production costs.

Note that all poultry take a break from laying each year to molt (grow new feathers to replace the old ones). You can expect the first molt at about 18 months of age. In addition, a hen will lay fewer eggs every year. It isn't economical to keep old laying hens after two or three years. If you aren't prepared to process them as stewing hens, your egg cost will increase.

If you want to gather eggs during winter, put a bright LED light on a timer in the coop. Set it to come on early in the morning and stay on for 14 to 15 hours a day. This will trick your chickens into thinking it's still summer. Ducks will often continue to lay during winter without a light.

There are additional considerations if you're raising poultry for meat production. Broiler chickens don't need as much space as laying hens, but ducks, geese, and turkeys all require more room. Check out the table in the Choose the Best Chickens and Other Poultry section. You'll need to process your meat birds at home or take them to a certified processing plant. Look for a local facility and ask about prices before you get started. Note that it is much more cost-effective to process your own birds.

The most popular broilers are Cornish hybrids because they gain weight rapidly and are ready to process at about eight weeks of age. Provide a minimum of 2 to 3 square feet of space in their coop (or more, if possible).

Broad-breasted turkeys and White Pekin ducks are also great for raising your own meat. The following table lists the costs of feeding meat birds, using the average cost for each expenditure shared in the lists Cost of 50 Pounds of Poultry Feed and Cost of Day-Old Poultry. Your total costs could look very different depending on waste, cost of baby poultry, feed, and other variables.

AVERAGE COST TO FEED MEAT BIRDS TO BUTCHER WEIGHT

POULTRY FOR MEAT	DAY-OLD COST	TIME TO BUTCHER	FEED TOTAL	FEED COST/LB.	MEAT YIELD	COST/LB. OF MEAT
CORNISH X ROCK BROILERS	$2.50	8 weeks	9.2 lbs.	$0.42	5 lbs.	$1.27
WHITE PEKIN DUCKS	$5	8 weeks	21.34 lbs.	$0.42	7 lbs.	$1.99
BROAD-BREASTED TURKEYS	$7.25	20 weeks	80 lbs.	$0.43	20 lbs.	$2.08

EXPECTED YIELD PER YEAR

Use the Expected Yield Per Year table in the HPAP template and the poultry production tables to estimate your yields of eggs and meat. Keep track of your actual yields when hens begin laying or meat birds are processed as well as the cost to produce those harvests.

ESTIMATED TIME

Use the following seasonal chores checklist to estimate the time commitment for your poultry projects. You'll need to provide care for your flock every day and find a farm sitter if you travel. There are also seasonal chores, and you'll have to raise replacement layers. Expect to spend about an hour each day feeding, watering, gathering eggs, mucking out the coop, letting the birds out in the morning, and putting them away at night.

SEASONAL CHORES FOR POULTRY CARE

Spring

- ↗ HATCH THE EGGS OR ORDER FROM A HATCHERY: 21 DAYS TO HATCH CHICKS
- ↗ SET UP THE BROODER: 30 MINUTES
- ↗ MOVE CHICKENS TO A DIFFERENT PASTURE: 1 HOUR
- ↗ CLEAN THE COOP: 1 HOUR
- ↗ RAISE REPLACEMENT LAYERS: 22 WEEKS
- ↗ RAISE BROILERS: 8 WEEKS

↗ PROCESS THE BROILERS: 30 MINUTES PER BIRD

Summer

↗ PROVIDE SHADE AND EXTRA WATER: 10+ MINUTES

↗ GATHER EGGS FREQUENTLY: 10+ MINUTES

Fall

↗ CLEAN THE COOP: 1 HOUR

↗ PROCESS TURKEYS: 30 MINUTES PER BIRD

↗ PUT A LIGHT ON A TIMER TO KEEP HENS LAYING: 10 MINUTES

↗ KEEP COOP VENTILATED: ONGOING

POTENTIAL PROBLEMS AND COST OF SOLUTIONS

Research any problems you might encounter when raising poultry and the cost of potential solutions. Predators, illness, and pests are common issues. Consider talking to your Cooperative Extension Service office or consulting local farmers to identify particular threats in your area.

For example, poultry is often lost to predators such as hawks, coyotes, and even dogs. Intestinal parasites, mites, Marek's disease, chronic respiratory disease, and sour crop are also common problems for laying hens. You can reduce parasites by rotating pastures and having chicks vaccinated at the hatchery.

IS YOUR POULTRY PROJECT FEASIBLE THIS YEAR?

After estimating the cost and time commitment for your poultry project as well as reviewing potential problems, is the project feasible in your first season of homesteading? If you feel that your project will take more time or a larger budget than you have available, can you break it down into smaller steps or spread it out over several years?

CREATING GOALS FOR POULTRY ACTION PLAN

If raising poultry is a feasible project for your homestead, it's time to set your goals. Make sure they align with your overall homestead goals. What do you want to accomplish during the first season? Do you want to keep a few laying hens to supply eggs so you can be more self-reliant? Are you planning to sell eggs at a farmers' market? Or are you hoping to raise chickens, turkeys, or ducks to fill your freezer for winter?

List your goals for keeping poultry this year, and break each one down into weekly and monthly step-by-step tasks. Use the breakdown tables of weekly, monthly, and first-year goals in the HPAP template to guide your plans. Include the estimated time and the cost of materials. Add the necessary chores to your homestead calendar to provide a comprehensive schedule. This will help you stay organized and prevent you from adding too much to your list.

Let's take a look at a weekly goal sample for preparing and caring for 25 chicks.

Sample Weekly Goal: Set up a brooder box, purchase 25 layer pullet chicks, and provide the proper care for the chicks.

DAY	TASKS TO COMPLETE	MATERIALS NEEDED	TIME	COST
MONDAY DATE: APRIL 1	✓ Create a brooder box for chicks	✓ Large plastic storage box ✓ Heat lamp and bulb	15 to 30 minutes	$25 $11
TUESDAY DATE: APRIL 2	✓ Purchase supplies and set up the brooder with a heat lamp	✓ Feeder ✓ Waterer ✓ Thermometer (have) ✓ 50 lbs. chick feed ✓ Probiotics ✓ Chick grit	30 minutes	$6 $6 $13.59 $3.69 $6.99
WEDNESDAY DATE: APRIL 3	✓ Check the brooder temperature ✓ Fill the waterer and feeder ✓ Line the brooder bottom with paper towels ✓ Purchase chicks ✓ Make sure the chicks are eating and drinking	✓ 25 layer pullets ($4 each)	2 hours	$100
THURSDAY DATE: APRIL 4	✓ Make sure the chicks are eating and drinking ✓ Clean brooder		15 minutes	$0
FRIDAY DATE: APRIL 5	✓ Check the temperature ✓ Clean the brooder		15 minutes	$0
SATURDAY DATE: APRIL 6	✓ Check the temperature ✓ Clean the brooder		15 minutes	$0
SUNDAY DATE: APRIL 7	✓ Check the temperature ✓ Clean the brooder		15 minutes	$0
		Totals:	4 hours	$172.27

PREPARE YOUR HOMESTEAD FOR POULTRY

Raising your own farm-fresh eggs and meat is a rewarding project. With a diet of greens, grass, and bugs, your hens will lay the most nutritious eggs possible. The yolks are deep gold and so much tastier than the pale ones from a commercial flock. You'll also know exactly how your poultry were raised and what they ate.

When laying hens have passed their most productive years, you can use them as stewing hens. Their bedding and manure make a wonderful addition to the compost pile to increase soil fertility for your vegetable garden. You can crush eggshells and feed them to your flock to increase their calcium.

There's a lot to consider when adding poultry to your homestead. The following sections break down the basics.

THE BEST SHELTERS

Your poultry need housing that will protect them from predators, rain, and wind. Make sure the coop is draft-free but has adequate ventilation to prevent a buildup of ammonia from their droppings. Provide a minimum of 2 square feet of coop space and 5 to 10 square feet of pasture for each chicken. They also need feed and water containers and a roosting bar with about 6 to 8 inches of space per hen. (The next section provides a table with the space required for other poultry.) You will also need one nesting box for every four hens. Allow space to store feed, bedding, and other supplies. Leave enough room to clean the coop and gather eggs.

A chicken tractor allows you to move a small flock to new grass each day. This is a great way to rotate their pasture, increase their nutrient intake, and save money.

THE BEST FEEDS

A sound diet for your chickens and other poultry is essential for their health and productivity. If your flock has access to pasture, they'll get nutrients from grass and insects, but you should also provide feed with a balanced formula. Here are the basic types:

Chick starter feed. Formulated for the needs of growing layer chicks or dual-purpose chicks, starter feed should provide balanced nutrition with at least 18 percent protein. Provide extra vitamin D for ducklings.

Layer feed. Laying hens need extra calcium and 16 percent protein. This is also a good option for ducks that are laying.

Meat producer feed. Broiler chicks need a protein content of 21 to 22 percent for best results. This formula is also a good option for reducing the cost of feeding broad-breasted turkeys after about two months of age through butcher weight.

All flock feed. All flock feed has 18 percent protein and is suitable for young pullets when they no longer need chick starter but aren't yet ready for layer feed. It is also good for ducks after the age of two weeks through butchering or until duck hens begin laying eggs. The calcium content is too low for layers.

Game bird feed. This high-protein (24 to 28 percent) poultry formula is best for turkeys, quail, and pheasant. This formula is not always available, so you may have to purchase meat producer feed and add supplemental protein.

Oyster shell. Keep this calcium supplement in a container separate from your laying flock's feed.

Grit. All poultry need some form of grit to help digest their food. Be sure to purchase a fine grit for chicks, as they can't swallow the larger grit.

CHOOSE THE BEST CHICKENS AND OTHER POULTRY

It's hard to resist baby ducklings, turkey poults, and guinea keets. There are also hundreds of different breeds of chickens, from small bantams to Brahmas the size of small turkeys. One of the biggest problems you might face is selecting just a few birds for your new homestead.

Before you order any poultry, make sure you are selecting the species and breeds that best fit your needs. If your goal is to raise an economical source of eggs, choose a hybrid or lightweight breed with low feed requirements. For the best meat production, raise broiler chicks, White Pekin ducks, or turkeys.

BEST POULTRY USES AND SPACE NEEDED

POULTRY SPECIES	USES	SPACE NEEDED
CHICKENS	Eggs, meat, heritage breeds	2 sq. ft. coop, 5 to 10 sq. ft. pasture
DUCKS	Eggs, meat, heritage breeds	3 to 4 sq. ft. coop, 15 sq. ft. pasture
GEESE	Eggs, meat, heritage breeds, eating weeds and grass	5 to 6 sq. ft. coop, 40 sq. ft. pasture
TURKEYS	Meat, heritage breeds	4 to 5 sq. ft. coop, 150 sq. ft. pasture

| GUINEA FOWL | Meat, eggs, insect control | Often allowed to roam freely for pest control |
| QUAIL | Meat, eggs, hunting | 1 sq. ft. of space |

BEST CHICKEN BREEDS FOR EGGS

When choosing layers, think about what is most important to you. Do you want the highest production for the lowest cost? Are customers in your area willing to pay a premium for a rainbow of egg colors? Or maybe you want to raise a dual-purpose breed that produces a decent yield of eggs and meat for your table.

The following table lists some of the best heritage breeds for egg production and dual-purpose breeds for both meat and eggs.

BEST LAYERS AND DUAL-PURPOSE BREEDS

HERITAGE CHICKEN BREEDS	PURPOSE/PRODUCTION/EGGS PER YEAR	SIZE (LBS.)	SPECIAL CHARACTERISTICS	COST OF CHICKS
AMERAUCANA OR EASTER EGGER	Layer/good/180 to 200	4 to 5	Colorful eggs: blue, green, pinkish, off-white	$3 to $5
AUSTRALORP	Dual-purpose/excellent/250	6.5 to 8.5	Brown eggs; calm but active	$3 to $5
BUCKEYE	Dual-purpose/good/120 to 150	6.5 to 9	Brown eggs; cold hardy; protective mothers	$5 to $6
DOMINIQUE	Dual-purpose/excellent/230 to 275	5 to 7	Brown eggs; does well in any climate	$3 to $4
HAMBURG	Layer/very good/200 to 225	4 to 5	White eggs; active; does well in any climate	$3 to $5

		SIZE (LBS.)		COST OF CHICKS
LEGHORN, WHITE	Layer/excellent/ 250 to 300	4.5 to 6	Economical egg production	$2 to $4
PLYMOUTH BARRED ROCK	Dual-purpose/very good/200	7.5 to 9.5	Brown eggs; cold hardy	$3 to $5
RHODE ISLAND RED	Dual-purpose/excellent/250+	6.5 to 8.5	Brown eggs; best heritage breed for brown egg production	$3 to $4
WYANDOTTE	Dual-purpose/good/180 to 200	6.5 to 8.5	Brown eggs; cold hardy	$4 to $5
HYBRID CHICKEN BREEDS	**PURPOSE/PRODUCTION/EGGS PER YEAR**	**SIZE (LBS.)**	**SPECIAL CHARACTERISTICS**	**COST OF CHICKS**
CINNAMON QUEEN	Dual-purpose/excellent/250 to 320	5 to 6	Extremely productive hybrid layers	$3 to $4
GOLDEN COMET	Dual-purpose/excellent/250 to 320	5 to 6	Extremely productive hybrid layers	$3 to $4

White Leghorns are one of the most economical heritage breeds of laying hens. Another option for an economical flock of laying hens are hybrids such as Golden Comets and Cinnamon Queens, which can produce 250 to 320 eggs per year. If you want laying hens that produce the most eggs for your investment, these are a great choice.

Money-Saving Tip: To keep your costs low, purchase pullets instead of straight run chicks from the hatchery. Straight run chicks will have both males and females and may seem like a good deal, but for meat production, cockerels are not as economical as broiler chicks.

BEST MEAT CLASSES

Poultry is one of the fastest ways to humanely raise your own meat on a homestead. The Average Cost to Feed Meat Birds to Butcher Weight table lists the average time to butcher, weight, and cost to produce broilers, ducks, and turkeys.

There are other options for meat chickens, such as hybrid Freedom Rangers or dual-purpose breeds. Be prepared to raise these birds for a longer period of time and harvest less meat from each one.

CHAPTER SEVEN

RAISING DAIRY AND MEAT ANIMALS

It's an enticing prospect to raise animals for meat and milk to complement your fresh eggs, fruits, and vegetables. Freshly made cheese and home-raised meats can save money, reduce waste, and instill the satisfaction of knowing you can provide for your family.

Of course, there are drawbacks to raising your own animal products. Regulations may limit your options, and each animal requires space, daily care, and shelter. It can be devastating to lose animals to disease or predators. Despite the potential difficulties, livestock can be a great addition for your homestead.

This chapter helps you create an action plan to assess the feasibility of raising dairy and meat animals on your homestead. It also provides information about preparing your property with shelter and fencing and choosing the best animals for your space. You can find easy recipes here that allow you to enjoy the dairy and meat you've raised.

CREATE A PROJECT ACTION PLAN FOR DAIRY AND MEAT ANIMALS

Dairy and meat animals can be great additions to a backyard homestead. There are many options to consider, even if you have a smaller space. For example, rabbits need very little room and provide meat, fiber, and furs. If milk production is one of your goals, goats are a great choice and can also be used for meat. Raising two or three pigs a year will fill your freezer, with extra to sell.

Before you dive into animal husbandry, work your way through the following information to make sure you are prepared for this investment. This chapter helps you create an action plan, set reasonable goals, and understand the basic costs and estimated time to carry out your plans.

Use the Homestead Project Action Plan (HPAP) template to create an action plan for your homestead. This section covers the specific considerations that will guide your research and assessment.

FEASIBILITY RESEARCH

Before you bring livestock home, you need to assess the time and financial investment involved in this project. This section helps you estimate the costs involved and determine the feasibility of keeping livestock in your backyard.

You'll also need to consider the daily and seasonal tasks related to keeping livestock, such as feeding, cleaning stalls, checking animals for signs of disease, and moving them to fresh pasture to reduce feed costs. Be realistic about how much time you have for this project.

It's also helpful to calculate how much you might save by raising your own animal products rather than buying them. Let's compare the cost of raising your own milk with buying it from a store, using the information in the Expected Cost of Materials section. We'll use goats to illustrate the math.

In 2019, the USDA reported that the average price of whole cow milk was $3.45 per gallon. According to the Cost Estimate for Purchasing and Feeding Dairy Animals table, it costs about $2 per day to feed one goat doe, and she will produce up to four quarts (one gallon) of milk daily. That's roughly $2 per gallon. When compared with $3.45 per gallon from the store, you could save $1.45 per gallon of home-raised milk.

If you have the capacity to produce and consume one gallon of milk for the 365 days of the year, you will save $529.25 annually. One gallon is more than most families need, but you can use the excess to make yogurt, cheese, and soap.

These calculations are very general, and they do not include the initial purchase price for a goat. When you are considering the feasibility of this project, keep in mind unexpected costs as well as ways to reduce expenses, such as selling offspring every year.

In your dairy and meat animals action plan, determine whether you have the budget and time to raise this livestock. Compare the expected cost of materials with your local prices, and determine what your budget allows. Read through the seasonal chores checklist to estimate your time commitment for this project.

EXPECTED COST OF MATERIALS

There are many expenses associated with raising livestock, and it's important to understand them before deciding whether this project is right for your homestead. Using this section's cost tables, your own research, and the Expected Cost of Materials table in the HPAP template, list all the materials you need to set up and maintain the project for the first year.

Let's take a closer look at the basics, such as feed and bedding. Check your local prices for the most accurate cost estimate, and use the following list as a general guideline. Make a list of necessary supplies and the cost of each item, then add up these expenses to calculate your total investment for raising livestock.

Livestock Basic Cost Estimate (50 pounds unless noted):

- Rabbit pellets: $15

- Goat feed: $20

- Goat minerals (25 pounds): $17

- Grass hay: $5 per 50-pound bale

- Cattle panel or hog panel: $25+

- Woven-wire fencing (47 inch by 330 foot roll): $289

- Electric fencing kit (100 feet): $370

- Fence posts: $3.50 each

- Rabbit hutch (36 by 24 by 34 inches): $60

- Livestock shed (12 by 8 feet): $1,000

Based on typical prices available online in March 2020.

Cost of Keeping Small Dairy Animals

To calculate the cost of feeding your dairy animals for a full year, look at the amount they consume each day and the cost per pound of feed. In the following cost estimate table, a Nigerian Dwarf doe eats three to five pounds of hay daily. One bale of hay weighs approximately 50 pounds, so one bale would feed one doe for about 10 days. If a 50-pound hay bale costs $5, that comes out to $0.50 of hay for one doe per day.

One dwarf doe also requires about one pound of goat rations each day. If 50 pounds of feed costs an average of $20, that comes out to $0.40 per pound daily. You'll also need to purchase about 25 pounds of minerals yearly for each doe for about $17. With 365 days in the year, that comes out to about $0.05 per day. When we add up the cost of hay, rations, and minerals, we get $0.95. In other words, it costs just shy of $1 each day to provide nutrition to one dwarf goat doe.

Using this method, we calculated the average cost of feed during milk production for dairy animals suitable for smaller homesteads. This table shows how much milk to expect each day while goats are lactating and the average cost of their feed.

COST ESTIMATE FOR PURCHASING AND FEEDING DAIRY ANIMALS

BREED/SPECIES	AVERAGE FEED COST PER DAY (IN ADDITION TO PASTURE)	PRODUCTION	COST AND SIZE OF LIVESTOCK
NIGERIAN DWARF GOAT	3 to 5 lbs. hay/day 1 lb. rations/day Free-choice minerals Average cost: $1	Up to 1.5 quarts/day of milk with 6% to 10% butterfat, which is good fresh or for making cheese and other dairy products; lactation lasts 9 to 10 months	$150 to $500 for a dairy-quality doe 75 lbs.
SAANEN GOAT	6 to 10 lbs. hay/day 2 lbs. rations/day Free-choice minerals Average cost: $2	Up to 4 quarts/day of milk with 3% to 4% butterfat, which is good fresh or for making yogurt; lactation lasts 9 to 10 months	$100 to $300 for a doeling 135 lbs.

Based on typical prices available online in March 2020.

These calculations are designed to give you a starting point for determining the cost of milk production. This table does not include the costs of housing, fencing, veterinary care, vaccinations, medications, or feeding animals when they are not producing milk.

Cost of Keeping Small Meat Animals

Rabbits are one of the most cost-effective meat animals to keep on a small property. If you have the space, you may also consider raising pigs or goats for the table.

When raising livestock for meat, you have to consider a few costs, including your initial investment in the animals, the amount of food they require to get to butchering age, and the cost of processing the meat.

Use the following table to determine which livestock to raise for meat. The investment to get started can be fairly low, as with rabbits. The cost to produce includes the purchase price of weaned pigs and the estimated cost of the feed necessary to raise them to butcher weight. The initial cost of breeding stock for rabbits and goats is not figured into the cost per pound to produce meat. The feeding cost includes the estimated amount needed to raise offspring to butcher weight.

Use this to help calculate your basic budget for raising meat at home.

COST ESTIMATE FOR PURCHASING AND FEEDING MEAT ANIMALS

LIVESTOCK	AVERAGE INITIAL COST OF LIVESTOCK	AVERAGE TOTAL FEED AND PROCESSING COST	AVERAGE PROCESSING FEE	AVERAGE MEAT YIELD	COST/LB. TO PRODUCE
	$20+ each for	2 (50-lb.) bags	Process meat at	10 fryers yielding	

RABBITS	breeding stock	($15 each): $30	home: $0	2.75 lbs. each for 27.5 lbs. total	$1.09
PIGS	$65 for a weaned pig	605 lbs.: $139.15	$232.25	152.25 lbs.	$2.87
GOATS	$150 to $300 for breeding stock	9.8 hay bales ($5 each): $49 98 lbs. goat rations: $39.20 Minerals: $4.90	$115	40 lbs.	$5.20

Based on typical prices available online in March 2020.

According to the table, you can expect a basic cost of $1.09 to $5.20 per pound to raise meat at home. This assumes you will butcher your rabbits yourself and does not include maintaining a buck.

Pigs produce a great deal of meat in a small area. Expect to spend anywhere from $50 to $80 for a weaned pig, or more for heritage breeds. Purchase them in spring and supplement their feed with garden leftovers over summer. They'll be ready for processing in fall.

In order for your pigs to put on weight, you'll need to provide grain in addition to pasture. Keep feed available at all times to achieve best weight gain. Butchering costs run $75 to $90 per pig, and the cost of cutting and wrapping is generally $0.95 to $1.25 per pound. Expect the dressed weight of a pig to be about 70 percent of the live weight, so a 217.5-pound animal would yield about 152.25 pounds of usable meat, as shown in the table here.

Goats have been gaining favor as meat animals in the United States. Expect a meat goat to reach a butcher weight of around 80 pounds. The ratio of usable meat to live weight varies, but plan on about 50 percent, or 40 pounds of meat for our example. If you have your goats butchered, expect processing to cost an average of $75 plus $1 per pound for cutting and wrapping, or $40. This brings the cost of processing to $115.

Here is the comparison of average prices of meat in 2020 to our estimated cost of home-raised meat:

MEAT TYPE	ESTIMATED PRICE OF HOME-RAISED MEAT PER LB.	AVERAGE COST PER LB. AT STORE	POTENTIAL SAVINGS PER LB.
RABBIT	$1.09	$8	$6.91
PORK	$2.87	$5	$2.31
GOAT	$5.20	$13	$7.80

Based on typical prices found online in March 2020.

We can calculate the average savings by adding the savings per pound for all three types of meat and dividing by three, which comes out to $5.67 in estimated savings per pound of home-raised meat.

If a family of four consumes an average of one pound of meat each day, this adds up to 365 pounds of meat per year. Taking into account the $5.67 in estimated savings, that works out to $2,069.55 in annual savings.

Here's what you could potentially raise on a small homestead each year:

♦ 2 rabbit does can each rear 6 litters of 10 kits, for a total of 120 rabbits weighing 2.5 to 3.5 pounds each, for an average of 360 pounds of meat.

♦ 2 pigs per season would provide an average of 435 pounds of meat.

♦ 2 goats per season would provide an average of 80 pounds of meat.

Use these cost estimates to narrow down your choices and determine which livestock fit your homestead goals. Think about how much space and time you can set aside for these projects. What meat will your family enjoy most? How much milk do you need to be self-reliant?

EXPECTED YIELD PER YEAR

Use the Expected Yield Per Year table in the HPAP template, along with information in the Choose the Best Dairy and Meat Animals section, to the determine the amount of milk or meat you could expect to harvest each year based on the number and breeds of animals you want to keep and can accommodate in your space.

ESTIMATED TIME

For an idea of how much time and energy are required to raise meat and milk, check out the following seasonal chores checklist. Your time will vary depending on how much livestock you have, as well as their individual needs. This checklist helps you determine whether you have enough time to raise and care for livestock without overwhelming your schedule.

SEASONAL CHORES FOR LIVESTOCK CARE

Spring

- ↗ PREPARE SUPPLIES FOR LIVESTOCK BIRTHS: 1+ HOUR
- ↗ ASSIST WITH BIRTHING: 1+ HOUR PER BIRTH
- ↗ ADMINISTER VACCINATIONS AND WORM MEDICATIONS: 15+ MINUTES PER HEAD
- ↗ BREED RABBITS FOR A SUMMER LITTER: 15+ MINUTES

- ↗ PURCHASE WEANED PIGS: VARIES
- ↗ MILK THE GOATS: 30+ MINUTES PER DOE

Summer

- ↗ WEAN YOUNG LIVESTOCK: 15+ MINUTES
- ↗ CHECK LIVESTOCK FOR PARASITES AND DISEASE: 15+ MINUTES
- ↗ ROTATE THE PASTURE AS NEEDED: 30+ MINUTES
- ↗ ORDER HAY AND STRAW: 1+ HOUR
- ↗ CHECK WATER FREQUENTLY IN HEAT: 15 MINUTES/DAY
- ↗ BREED RABBITS: 15+ MINUTES
- ↗ BUTCHER THE SPRING LITTER OF RABBITS AS FRYERS: 30 MINUTES/HEAD
- ↗ MILK THE GOATS: 15+ MINUTES PER DOE

Fall

- ↗ SET UP LIVESTOCK PROCESSING WITH A BUTCHER: 15+ MINUTES
- ↗ STOCK UP ON HAY AND BEDDING FOR WINTER: 1+ HOUR
- ↗ BREED RABBITS: 15+ MINUTES
- ↗ BREED GOATS IN FALL OR WINTER: VARIES
- ↗ CLEAN OUT THE BARN: 2+ HOURS

Winter

- ↗ CHECK WATER BUCKETS AND KEEP THAWED: 15 MINUTES/DAY
- ↗ CLEAN THE STALLS: 20+ MINUTES/DAY
- ↗ BUTCHER THE RABBITS: 30 MINUTES/HEAD
- ↗ BREED RABBITS: 15+ MINUTES
- ↗ TOTAL LIVESTOCK EXPENSES AND DETERMINE BUDGET FOR NEXT SEASON: 1+ HOUR

POTENTIAL PROBLEMS AND COST OF SOLUTIONS

Research any problems you might encounter when raising dairy and meat animals, as well as potential solutions and their costs. Common issues include mastitis in dairy animals, the time commitment for milking and butchering, and loss of livestock to disease or predator attacks. Vaccinations and worming are very important in preventing disease in livestock.

IS YOUR MEAT AND MILK PROJECT FEASIBLE THIS YEAR?

After reading through the seasonal chores and cost estimates, do you have the budget and time to raise livestock successfully? If not, are there ways you could modify your plans and keep a smaller number of animals for the first season or spread your projects out over several years? Use the HPAP template to record your ideas.

CREATING GOALS FOR YOUR DAIRY AND MEAT ANIMALS ACTION PLAN

What are your goals for raising dairy and meat animals? Do you want to increase self-reliance or raise a product for sale? Be sure to research the regulations for the sale of milk and meat in your area before planning a side business. Set goals for your first year, then break them down into weekly and monthly goals. Use the breakdown tables of weekly, monthly, and first-year goals in the HPAP template to guide your plans. Include the estimated time and cost of materials, and add the necessary chores to your homestead calendar.

Let's take a look at a sample weekly goal for setting up an area for breeding and raising rabbits for meat.

Sample Weekly Goal: Set up an area for the rabbits, assemble the hutches, and purchase a breeding trio of New Zealand White rabbits.

DAY	TASKS TO COMPLETE	MATERIALS NEEDED	TIME	COST
MONDAY DATE: MARCH 20	⟋ Clear an area for 3 rabbit hutches ⟋ Purchase the hutches, feed, bedding, water bottles, and feeders	⟋ Hutches ⟋ 50 lbs. of feed ⟋ Bale of straw ⟋ Water bottles × 3 ⟋ Feeders × 3	1 hour 30 minutes	$180 $15 $5 $78 $24
TUESDAY DATE: MARCH 21	⟋ Assemble the rabbit hutches ⟋ Order a breeding trio of New Zealand Whites	⟋ Tool included in the kit	1 hour 30 minutes	
WEDNESDAY	⟋ Store the rabbit feed in a metal garbage can in the chicken coop			

DATE: MARCH 22	✓ Add straw to the hutch shelters ✓ Fill water bottles, plug them in, and test them	✓ Outdoor extension cords and power strip (have)	30 minutes	
THURSDAY DATE: MARCH 23	✓ Pick up rabbits from the breeder ✓ Fill the rabbit feeders	✓ 3 pet carriers (have) ✓ Feeders and feed	1 hour	$60
FRIDAY DATE: MARCH 24	✓ Daily care: Check the feed and water and clean cages		30 minutes	
SATURDAY DATE: MARCH 25	✓ Daily care		30 minutes	
SUNDAY DATE: MARCH 26	✓ Daily care ✓ Wait 2 weeks for rabbits to acclimate before breeding		30 minutes	
		Totals:	6 hours	$362

PREPARE YOUR HOMESTEAD FOR DAIRY AND MEAT ANIMALS

Farm animals need safe shelters and strong fencing to prevent them from escaping and to keep out predators. You don't want to lose your investment or waste time rounding up wayward livestock.

Animal shelters also need additional space for storing feed and supplies. If you plan to breed livestock, you'll have to include room for their offspring. Living quarters are a big investment, so consider building your own.

Consult your homestead map and look at the areas set aside for raising livestock. Is there shade in summer? Will they be protected from strong winds in winter?

Plan so you have enough time to purchase fencing, housing, and other supplies when they are on sale or locate decent used materials. If you intend to build from scratch, give yourself time to complete these projects before the animals arrive.

FENCING

There are a variety of options for fencing in your animal pens. Choose the one best suited to your livestock, budget, and property.

Cattle and hog panels are fairly expensive, but they require less maintenance once installed. They generally come in 16-foot sections and have openings between wires that are large enough for horned goats to get their heads stuck in the holes. To prevent this, attach chicken wire to the inside of the panels or install electric fencing inside the panels.

Woven-wire fencing comes in rolls in a variety of heights and lengths, so you can choose the size you need. The wires are generally set closer together, which prevents livestock from sticking their heads through the holes. There are different lengths and heights available.

Electric netting fences made for portability are a great option for pasturing livestock that are prone to escape. Pigs like to root under traditional fences, and goats like to climb over them, but electric netting fences work well to contain both animals. You'll need to keep grass trimmed to prevent it from touching the bottom wire and shorting out the fence. You'll also have to buy fiberglass posts and a solar-powered fence energizer.

Pallet fences are a popular way to fence in small livestock and might be useful for homesteaders on a limited budget.

SHELTER

Take care when choosing the location for permanent animal shelters. Make sure buildings have adequate ventilation but are free from drafts that can cause respiratory illness. You need a shelter that is large enough to house all the necessary supplies and includes room for cleaning out manure, milking goats, and housing breeding stock.

You can keep pigs and goats in open housing that protects them from the sun, rain, and wind. In cold climates, goats may need closed housing to keep them healthy. Pigs are fairly happy in a pen of about 100 square feet per animal. Goats need at least 10 square feet of bedded area and 25 square feet of pasture in an open housing arrangement. In a closed barn, provide a minimum of 6 square feet of stall for dwarf breeds plus 15 square feet of pen per animal. Allow space for offspring, a milking stand, feed, and other supplies. If you intend to keep a buck for breeding, he'll need his own pen and stall to prevent unplanned kids.

Housing rabbits for home meat or fiber production is inexpensive and space efficient. If you are housing rabbits in a barn or a shed, you can keep them in wire cages. You can also keep rabbits in hutches outdoors. Each hutch should have an enclosed box that provides shelter in rain and wind. Rabbits can overheat, so shade in summer is more important than protection from cold in winter. Some homesteaders raise their rabbits in movable enclosures, similar to chicken tractors, so they can graze. You can build these out of scrap lumber, pallets, or other materials and create a wire bottom to prevent escape.

Each breeding doe and buck needs a separate cage or hutch. Does ready to kindle (give birth) need a nesting box and enough space for their litter to reach weaning age. Provide a hanging feed dish and water bottle that is heated in winter to prevent freezing.

CHOOSE THE BEST DAIRY AND MEAT ANIMALS

To make the most of your homestead, be sure to choose animals that fit your space and budget while providing the goods you want. Certain animals can fill more than one need. For example, some rabbits provide both meat and fiber, and a mixed herd of goats can produce milk, meat, and wool. If you choose the best species or combinations of breeds, you can maximize the harvests from your livestock and maybe even sell young stock at a profit. Let's take a closer look at some of the best choices for backyard homesteads.

THE BEST ANIMALS FOR DAIRY

Providing your own milk for fresh use or making dairy products such as cheese is an important goal for many homesteaders hoping to increase their self-sufficiency. Goats are the best dairy animals for small homesteads, and even an urban homesteader might have space for two Nigerian Dwarf goats.

Goats provide meat, milk, and maybe even wool for spinning your own yarn. They are also great at clearing brush and can be used as pack animals. Although you can keep all goats in fairly tight quarters, the average backyard homestead is best suited to Nigerian Dwarf goats to provide milk and possibly meat or an income from selling kids. If you have more room, you could keep Saanens or another full-size breed.

Good-quality hay made up of less than half alfalfa or clover provides enough nutrition for adult goats that are not pregnant or lactating. Expect a full-size goat to consume two to five pounds of hay each day. You'll have to provide goat rations to growing kids and pregnant or lactating does. Supply free-choice goat minerals or a salt block formulated with the micronutrients they need.

Never change your goats' diet suddenly or allow them to overeat and get bloated, which can be fatal (free-choice baking soda can help treat this ailment). Goats are susceptible to parasites and a number of diseases, so use worming medications and vaccinate regularly. Rotate their pasture each year to reduce parasite load. You might want to have kids disbudded so their horns don't cause safety issues for you and your goats.

Tips for Keeping Goats: Goats love to eat brush and can clear an overgrown area. Be careful of rough undergrowth, however, as it can damage the large udders on dairy goats. Goats are herd animals, so keep at least two.

THE BEST ANIMALS FOR MEAT

Raising livestock for meat can be difficult. It's easy to get attached to animals and feel sad about processing them. But when you raise your own livestock, you know they have been treated well, and you're also aware of what they consumed, which eases your own health concerns.

In an average-size backyard homestead, the easiest and most space-efficient livestock for meat production are rabbits, goats, and pigs. If you have one or more acres to raise sheep and Dexter cattle, they also provide quality lamb and beef in addition to wool and milk.

Rabbits

One buck and two does can provide a great deal of meat, and they don't require much feed or a large investment to get started. Expect to spend $20 or more for each buck and doe to start your operation.

Some of the best breeds of rabbit for meat production are New Zealand White, Californian, Champagne d'Argent, Flemish Giant, and English Spot. The New Zealand White is commonly available and hardy.

Tips for Raising Rabbits: Provide free-choice timothy hay for rabbits. Never change their diets suddenly because doing so can cause fatal diarrhea. Introduce new feed or treats gradually. A doe can be bred six times a year and give birth to 8 to 12 kits in each litter. Does are territorial, so always bring the doe to the buck for breeding. A doe will attack any rabbit placed in her hutch.

Pigs

You can purchase two or three young pigs in spring or early summer and take them to a custom butcher shop in fall for processing. Some heritage breeds provide lean meat, such as Guinea Hogs or Tamworth, but these breeds are more expensive and harder to find. Common breeds such as Duroc or Berkshire are more readily available and grow rapidly.

Tips for Keeping Pigs: Pigs love to dig up grubs, rodents, and roots. Move your pigs to a field where beets, grains, pumpkins, or other crops have been harvested and let them search for any leftovers. This will supplement their feed, and they'll also help prepare the ground for next year's crop. Pigs need shade to prevent sunburn.

Goats

Although you can harvest any goat for its meat, you will get a higher yield from a Boer, Kiko, or other meat breed. A Boer goat can be butchered at seven to eight months of age. Not all butcher shops will process goats, so do your research ahead of time. Expect a purchase price of $200 or more for each doe and buck to start your herd.

One option is to keep dairy goats and breed them with a Boer buck. The offspring will have the meat production of the Boer, but your dairy does will produce a decent amount of milk.

CHAPTER EIGHT

BEEKEEPING FOR BEGINNERS

Keeping bees is very beneficial for your backyard homestead. A healthy colony will pollinate your orchard, herbs, vegetable garden, and flowers and increase your harvests. In addition, bees work tirelessly to collect nectar and produce honey, one of the most delicious homegrown products for your table. Beeswax, the by-product of honey collection, is wonderful for making your own salves, balms, and candles. You can even rent out honeybees to pollinate surrounding orchards. It's amazing that such small creatures can provide so many services.

This chapter provides the information you need to decide whether a beekeeping project is right for your homestead. We'll discuss how to prepare your homestead for bees by eliminating toxins and providing food sources. You'll learn how to determine the best location on your property for raising bees, choose the best hive, and install and harvest your hive. For those of you with a sweet tooth, there are some delicious recipes for honey here plus instructions on making your own balms and salves with beeswax.

CREATE AN ACTION PLAN FOR YOUR BEES

If beekeeping is part of your homestead master plan, you're not alone. There's a renewed interest in this microlivestock as a way to boost crop production and increase self-reliance. You can keep several hives on an average-size lot, so beekeeping and honey sales are a potential side business for almost anyone.

Use the Homestead Project Action Plan (HPAP) template to create a beekeeping action plan for your backyard homestead. This section covers the specific considerations that will guide your research and assessment.

FEASIBILITY RESEARCH

Before you place an order for beekeeping equipment, do some research to find out if this project is right for you. Bees are fairly inexpensive and require less time than most livestock, but you still have to invest in equipment, perform regular inspections, and feed at critical times to help them thrive.

Jot down your reasons for adding bees to the homestead. Do you want bees to pollinate crops for better production? Are you hoping to harvest the honey and beeswax? The information in the following sections will help you understand the costs associated with your beekeeping goals as well as the time involved for you to be successful.

Begin by looking at how much you might save by raising your own honey rather than purchasing it. Let's compare the cost of raising your own honey with buying it from a store, using the information found in the Expected Cost of Materials section. Most of the equipment you purchase should last at least five years, so we'll use an average yearly expense of $112.30. Assuming you harvest a somewhat modest 40 pounds of honey each year and use the extractor and tools from a beekeeping club, your cost will be about $2.81 per pound.

Quality honey costs $8 to $9 per pound, so we'll use the average cost of $8.50 per pound to compare. If your honey costs $2.81 per pound, you're saving $5.69 per pound.

If your family uses one pound of honey each week, your total consumption would be 52 pounds annually, or a potential annual savings of $295.88.

Although you don't need honey to survive, having your own natural sweetener will certainly make life more enjoyable. You'll also reduce household waste and increase your self-reliance. Be aware, however, that your harvests will vary and you may not harvest any honey from your hive during the first year.

In your beekeeping action plan, determine whether you have the budget and time to raise bees successfully. Compare the expected cost of materials with current prices and determine what your budget allows. Spread out expenses over several years if you don't have enough funds to cover the entire project right away. Read through the seasonal chores checklist to calculate the approximate time commitment for this project, and be realistic about whether you can take it on.

EXPECTED COST OF MATERIALS

You'll need some supplies to get started in your beekeeping enterprise. The following list is geared toward beekeepers choosing a Langstroth hive. Using this section's cost lists, your own research, and the Expected Cost of Materials table in the HPAP template, list all materials you'll need to set up and maintain the project for the first year. Note that you don't have to purchase all this equipment right away. Essentials are marked with an asterisk (*).

Beekeeping Basic Equipment Costs:

- Beehive starter kit*: $200+

- Varroa screen: $3.60+

- Brood box and frames: $16+

- Supers and frames: $20+ each

- Queen excluder: $5+

- Foundation: $4+

- Nucleus hive: $200+

- Packaged bees: $135+

- Beekeeping suit: $110+

- Hat and veil*: $20+

- Gloves: $20+

- Smoker*: $15+

- Hive tool*: $10+

- Bee feeders: $7+

- Extractor: $80+

- Uncapping knife: $25+

Here's a breakdown of the basics for a Langstroth hive:

- Starter beehive: $200 to $300

- ◆ Extra supers and brood boxes for the first year: $60 to $100

- ◆ Hat and veil: $20 to $40

- ◆ Smoker: $15 to $50

- ◆ Hive tool: $10 to $13

- ◆ Bees: $135 to $180

Your initial investment comes to $440 to $683, or an average cost of $561.50. These supplies will serve you well for at least five years, and although you are sure to need more supplies, you could get by for some time, especially if you build your own hive components.

EXPECTED YIELD PER YEAR

Use the Expected Yield Per Year table to record the amount of honey you might expect from your beekeeping project. Assume lower yields from beehive systems other than a Langstroth hive. You may not harvest honey in the first year, but you should see an increase in pollination and yields from many crops.

ESTIMATED TIME

Your beehive needs only a weekly inspection during the busiest time of year, when the nectar flow peaks. Use the following checklist to determine the feasibility of this project, create your action plan, set reasonable goals, and fill in your homestead master calendar.

SEASONAL CHORES FOR BEEKEEPING

Spring

- ↗ FEED THE BEES 1:1 SUGAR WATER WHEN NECTAR IS SCARCE: 15 MINUTES

- ↗ INSPECT THE HIVE FOR DISEASE AND PARASITES: 30+ MINUTES

- ↗ ADD NEW HIVES: 1+ HOUR

- ↗ ADD BENEFICIAL PLANTS TO THE GARDEN: 30+ MINUTES

Summer

- ↗ CHECK SUPERS WEEKLY AND ADD AS NEEDED: 30+ MINUTES

- ↗ PROVIDE A SHALLOW PAN OF WATER FOR BEES: 5 MINUTES

- ↗ CATCH AND HOUSE SWARMS: 1+ HOUR

- ↗ CHECK FREQUENTLY FOR DISEASE AND PESTS AND TREAT AS NEEDED: 30+ MINUTES

- ↗ HARVEST HONEY AS NEEDED: 2+ HOURS

Fall

- ↗ INSPECT THE HIVES, CHECK FOR DISEASE AND PESTS AND TREAT THEM: 30+ MINUTES

- ↗ HARVEST SURPLUS HONEY: 2+ HOURS

- ↗ FEED 2:1 SUGAR WATER IN LATE FALL: 15 MINUTES

- ↗ CLEAN THE EQUIPMENT AND USED BROOD BOXES AND SUPERS: 1+ HOUR

Winter

- ↗ ORDER SUPPLIES: 15+ MINUTES
- ↗ BUILD NEW HIVE COMPONENTS: VARIES
- ↗ PLAN FOR UPCOMING SEASON AND REVIEW NOTES: 1+ HOUR
- ↗ KEEP SNOW CLEARED FROM THE HIVE ENTRANCE: 15+ MINUTES
- ↗ CHECK THE HIVES ON WARM DAYS: 15+ MINUTES

POTENTIAL PROBLEMS AND COST OF SOLUTIONS

Research any problems you might encounter when keeping bees, as well as possible solutions and their associated costs. Common issues are varroa mites, American and European foulbrood disease, and colony collapse disorder, as well as mice and bears.

IS YOUR BEEKEEPING PROJECT FEASIBLE THIS YEAR?

After reading through the seasonal chores checklist and the cost estimates, do you have the time and budget to add beehives to your homestead this year?

Are there ways you could modify your plans and build a top-bar hive instead of purchasing a Langstroth system, or can you spread out the expenses over several years? Jot down your ideas.

CREATING GOALS FOR BEEKEEPING ACTION PLAN

If raising bees is a feasible project for your homestead, it's time to set some goals. List your goals for keeping bees this year, and break down each one into the weekly and monthly step-by-step tasks. Use the breakdown tables of weekly, monthly, and first-year goals in the HPAP template to guide your plans. Include the estimated time and the cost of materials needed to reach your goals this year. Add the necessary chores to your homestead calendar to provide a comprehensive schedule.

Let's look at a sample weekly goal of setting up a beehive.

Sample Weekly Goal: Set up beehive, bring home new nucleus hive, and transfer the colony to the starter hive.

DAY	TASKS TO COMPLETE	MATERIALS NEEDED	TIME	COST
MONDAY DATE: APRIL 20	✗ Prepare space for the beehives ✗ Assemble the starter hive and check tools ✗ Plant nectar-producing flowers	✗ Beehive starter kit ✗ Wildflower seeds	4 hours	$200 $3
TUESDAY DATE: APRIL 21	✗ Pick up the nucleus hive ✗ Move the starter hive and replace with nucleus hive ✗ Open nucleus hive	✗ Nucleus hive ✗ Beekeeping suit and veil ✗ Sugar water	1 hour 30 minutes	$200 $110 $20
WEDNESDAY DATE: APRIL 22	✗ Watch the bees ✗ Water the wildflower seeds		15 minutes	
THURSDAY	✗ Watch the bees		15	

			minutes	
DATE: APRIL 23	↗ Water the wildflower seeds		minutes	
FRIDAY DATE: APRIL 24	↗ Watch the bees ↗ Test the smoker	↗ Smoker and cartridge	15 minutes	$18
SATURDAY DATE: APRIL 25	↗ Transfer the colony to a starter hive and fill the feeder with sugar water ↗ Water the wildflower seeds	↗ Starter hive ↗ Beekeeping suit and veil ↗ Smoker and cartridge ↗ Hive tool	1 hour 15 minutes	$10
SUNDAY DATE: APRIL 26	↗ Watch the bees coming and going from the starter hive		15 minutes	
	Totals:		7 hours 45 minutes	$561

PREPARE YOUR HOMESTEAD FOR BEES

Raising your own honeybees is a pretty sweet job, but there's a lot to do before you can harvest their delectable bounty. Join a local beekeeping club or find a mentor to show you the ropes. It's helpful to have someone experienced on hand for your first hive inspection.

Being a part of a club is also a great way to learn. If you join one in your area, you may be able to use a club extractor and other tools when your supers start filling up with honey. In addition, beekeeping clubs sometimes capture bee swarms. Put your name on a waiting list to receive a free swarm.

Prepare for bees by planting sources of nectar and pollen. Eliminate all pesticides sprayed on lawns and use organic sprays when bees are inactive. Dandelions and other so-called weeds are a great source of food. Bees need nectar most in early spring and late fall, when fewer plants are in bloom.

Honeybees are important pollinators of vegetables, herbs, fruits, nuts, and some field crops, such as buckwheat and clover.

The following list of nectar-producing plants is a good place to start when choosing crops and flowers for your bees. Plant large blocks or beds to provide more nectar for bees and plan a succession of flowers for a continual food source. Keep in mind that this list is not exhaustive, and some plants will do better than others in your zone. Research before purchasing any plants.

PLANTS THAT PROVIDE NECTAR FOR HONEYBEES

SPRING	SUMMER	FALL
Alyssum	Alfalfa	Anemone
Black locust	Bee balm	Aster
Bleeding heart	Berries	Bugbane
Chickweed	Black-eyed Susan	Bush clover
Cornelian cherry	Buckwheat	Calendula
Crocus	Calendula	Caryopteris
Currants	Clover	Coneflowers
Dandelions	Cosmos	English ivy
Daphne	Gaillardia	Goldenrod
Fruit trees	Goldenrod	Helianthus
Gooseberry	Herbs	Joe-pye weed
Hazelnut	Hyssop	Lemon balm
Heather	Lavender	Michaelmas daisy
Hellebore	Liatris	Pansy
Jonquils	Marigolds	Russian sage
Lilacs	Milkweed	Sedum
Lungwort	Mint	Snapdragon
Magnolia	Nasturtium	Witch hazel (common)
Maple trees	Peony	Zinnias
Muscari	Phlox	
Pansies	Salvia	
Rockcress	Sedum	
Spicebush	Sunflowers	
Tulips	Vegetables	
Violets	Vetch	
Witch hazel (vernal)	Zinnias	

FINDING THE BEST SPACE FOR YOUR BEES

You don't need a lot of space to keep bees as long as there are plenty of sources for nectar and pollen within three miles of the hive. Even urban homesteaders can raise a colony on a rooftop, balcony, or small patio. My sister in New York City purchases locally raised rooftop honey at her neighborhood farmers' market. If you have at least 2 square feet for a hive plus space to move around it, you have enough room for bees.

Select a location that is protected from wind and gets morning sun. Leave about 3 feet around the hive for easy access, plus 5 to 10 feet of runway for bees leaving the entrance. Honeybees do their business as they leave the hive, so don't plant your vegetables or nectar-producing plants in this area.

Avoid situating the hive close to a public area, walkway, or playground. Leave 3 feet between hives and make sure the entrance of each hive faces a slightly different direction, or paint each one a different color so the bees can tell them apart. You also need a clean, dry place to store your supplies that provides convenient access to the hives.

CHOOSING THE BEST HIVE

Several beehives work well for various goals and management styles. Here are the three most commonly used systems in the United States, as well as the pros and cons of each.

Langstroth Beehive

This is the most widely used design and produces the most honey. Langstroth hives are a series of stacked boxes loaded with frames for bees to raise brood (eggs and larvae) or store honey. In this system, beekeepers add new brood boxes and supers as the existing ones fill, which gives the colony room to grow. The brood boxes are placed on the bottom, and the supers for honey are stacked on top. A queen excluder is often placed between the brood and supers to prevent larvae in the honeycomb.

The frames in each box are strategically placed to provide the proper bee space (the room bees need to work naturally). Each frame has four sides that hold in place a foundation, or a sheet of wax or plastic. The frames and foundation increase your costs and make it a bit trickier to build your own hive components.

The Langstroth hive typically has a flat cover with ventilation to reduce moisture, but it is subject to condensation over winter. A screen is placed over the floor of the hive to trap Varroa mites. The floor also has adjustable entrance blocks.

Warre Beehive

The Warre beehive system may be best for those who want to use more natural beekeeping methods or have problems with moisture levels. The Warre hive looks similar to the Langstroth hive, but the components are smaller and the roof has a peak in the center with sloped sides. It also allows for increased ventilation and has a built-in moisture control system called a quilt box.

The Warre hive is built to mimic a hollow tree, a favored location for wild bees. The brood boxes and supers have bars across the top of each component without a frame or foundation, which allow the bees to build their comb more naturally. The brood box is placed on top, so it's easier to inspect the developing larvae and queen. However, when you need to place new supers on the bottom or harvest honey, you have to remove the top boxes.

Because there are no frames or foundation for the comb, honey is harvested without an extractor (smash the comb and strain manually) or the comb is placed in a wire basket before loading into the extractor.

Top-Bar Hive

The top-bar hive (TBH) is easy to build and maintain. This design is gaining in popularity among people who want to increase pollination, keep bees naturally, and harvest honey for personal use.

The TBH is constructed as one long box that has a smaller bottom and a larger top to mimic the natural shape of honeycomb. The box is topped with a peaked roof that may be hinged or lifted off; it also allows rain to run off. Instead of frames and foundation, the bees build their comb hanging from evenly spaced bars across the top of the box. Some TBH designs have two compartments with a divider to exclude the queen from the honeycomb. New bars are added across the top to accommodate more brood or more honey as needed.

The TBH is normally raised, so you don't have to bend over while working. There are no modular brood boxes or supers involved in this design, and you harvest the honeycomb by lifting and removing the individual bars of comb. This makes the TBH system much easier for people who can't lift heavy supers.

The following table covers the pros and cons of each type of beehive. Consider the method of beekeeping you want to use and the best hive for your purposes. Use this information to set your goals and work on your step-by-step action plan for keeping bees.

COMMON BEEHIVES AND THEIR PROS AND CONS

BEEHIVE TYPE	PROS	CONS
LANGSTROTH	Maximum honey production in a small space Expandable, modular system Queen excluder may be used to prevent eggs and brood in supers Standard sizes for brood boxes and supers mean the components are usually compatible between manufacturers Frames with a foundation can be put in extractor for easy harvest	Not designed according to bees' natural habitat Moisture can build up in the top of the hive over winter, when warmth of the bees causes condensation Frames and foundation can be expensive The brood box is at the bottom, so inspecting the brood is more laborious Components are heavy and cumbersome
WARRE	Expandable, modular boxes are smaller and lighter than Langstroth boxes Allows bees to build a comb naturally Brood boxes are on top for easy inspection Boxes have a slat across the top for bees to build comb naturally instead of full frames with a foundation Less expensive and easier to build Top of the hive (the quilt box) has an absorbent material to wick the moisture created by the bees in winter	Boxes are added to the bottom of the stack instead of the top, creating more labor Must use a wire basket to place the comb in an extractor or destroy the comb to harvest honey Hive components can be heavy and cumbersome compared with top-bar hive
TOP-BAR HIVE	The hive is situated at a comfortable height for the beekeeper Allows bees to build comb naturally No boxes to lift off, so it's easier to open and inspect the brood; frames are lifted out to inspect and harvest No foundation, so it is less expensive Easier to build yourself	Must use a wire basket to place the comb in an extractor or destroy the comb to harvest honey Components aren't a standard size Bees may get crowded more quickly and swarm because of limited expansion capacity Some designs do not allow for excluding the queen from laying eggs in honeycomb

INSTALLING AND MAINTAINING YOUR BEEHIVE

Before you bring your bees home, complete the following steps, which will vary depending on whether you are purchasing a working hive, packaged bees, or a nucleus hive (kept temporarily in a nuc box):

1. Build or order your starter hive and components.

2. Clear and level the area where you will place your hive.

3. Set up the hive.

4. Treat components with beeswax or paint the exterior of the hive.

Bringing Your Nuc Box Home

When you are ready to bring your bees home, follow these steps:

1. Bring the bees home.

2. Move the beehive from its stand.

3. Set the nuc box on the floor with the entrance facing the direction the permanent hive will face.

4. Open the nuc box entrance.

5. Leave the nuc box in place for at least 24 hours, but preferably 3 or 4 days.

Transferring Your Bees

When you are ready to transfer your bees to the hive, make sure you have the following equipment:

♦ Brood box

♦ Empty super

- Crown board

- Empty deep frames for brood box

- Roof

- Bee feeder with 1:1 sugar water solution

- Bee suit, hat and veil, gloves

- Smoker

- Hive tool

Follow these steps to transfer your bees from the nuc box to the hive.

1. Transfer your bees on a warm, sunny day.

2. Light your smoker and put on your bee suit, gloves, and veil.

3. Stand next to or behind the nuc box and direct a little smoke at the entrance.

4. Wait a few moments, then move the nuc box a few feet away from the hive base.

5. Place the brood box on the floor of the hive and put the entrance block in place.

6. Place one or two frames with foundation in the brood box at the outer edge of the box.

7. Place the nuc box next to the hive and use the hive tool to gently pry off the top.

8. Give the nuc box a gentle puff of smoke.

9. Gently pry one of the end frames out of the nuc box.

10. Place the frame in the brood box next to the empty frame with foundation, with the same side facing out as before.

11. Remove each frame and place it in the brood box, keeping the frames in the same order and position.

12. Be careful not to damage the queen.

13. Fill the brood box with empty frames with foundation.

14. Shake any remaining bees into the brood box.

15. Turn the nuc box upside down and lean it against the hive over the entrance so any remaining bees can find their way in.

16. Fit the cover board on top and make sure it is level.

17. Set the empty super on top of the cover board to make room for the feeder.

18. Place the feeder over the hole in the crown board.

19. Place the roof on top and make sure it is secure.

20. Don't open the brood box to inspect the bees again for about a week, but keep the feeder filled.

21. Continue feeding the bees with sugar water until they have drawn out the frames in the brooder box.

Maintaining Your Hive

Although bees don't need a lot of care, you have to perform these tasks on a regular basis to keep them healthy and prevent them from swarming:

- ↗ INSPECT SUPERS REGULARLY DURING SUMMER AND ADD NEW SUPERS AS THE EXISTING ONES FILL WITH HONEY.

- ↗ INSPECT BROOD BOXES AND ADD A NEW ONE WHEN THE EXISTING BOX IS NEARLY FILLED. IF BEES FEEL LIKE THEY ARE RUNNING OUT OF SPACE, THEY MAY SWARM TO LOOK FOR A NEW HOME.

- ↗ CHECK FREQUENTLY FOR SIGNS OF DISEASE AND PARASITES AND TREAT YOUR HIVE AS NEEDED.

- ↗ CLEAN YOUR EQUIPMENT AND HIVE COMPONENTS THOROUGHLY TO PREVENT DISEASE AND PARASITES.

- ↗ KEEP EXTRA BROOD BOXES AND SUPERS ON HAND DURING SPRING AND SUMMER TO EXPAND THE HIVE AS NEEDED.

- ↗ WATCH FOR SIGNS THAT THE QUEEN IS NO LONGER PRODUCTIVE AND CHECK FOR QUEEN CELLS IN THE BROOD FRAMES (ENLARGED CELLS).

- ↗ ORDER A NEW QUEEN IF YOUR HIVE DOES NOT HAVE A PRODUCTIVE QUEEN. MAKE SURE THE WORKERS ACCEPT HER BEFORE RELEASING HER FROM HER CAGE.

- ↗ FEED BEES SUGAR WATER DURING PERIODS OF LOW NECTAR PRODUCTION.

- ↗ REMOVE HONEY ONLY IF THERE IS ENOUGH TO FEED BEES DURING LOW NECTAR FLOW AND OVER WINTER.

Harvesting Honey

Harvest honey in summer or early fall, when there is a surplus to feed the bees and your sweet tooth. Remove frames of capped honey from supers and bring them into a closed area where bees won't get in the way of safe processing.

Follow these steps:

1. Cover surfaces with plastic or newspaper.

2. Use an uncapping knife or fork to remove caps from honey frames.

3. Place frames in an extractor and spin to remove honey from the comb.

4. Strain the honey through sieves to remove any wax or bee parts.

5. Allow the honey to sit until bubbles rise to the top.

6. Fill clean jars with honey and cap with sealed lids to keep moisture out.

7. Label the jars with the date, hive, and main source of nectar.

Tips for Beginners: The smell of smoke encourages honeybees to eat honey in preparation to flee a fire instead of attacking you. Don't wear dark-colored clothing around your hives—the colony may mistake you for a bear, their natural enemy. Remain calm and move slowly when working with your bees to prevent upsetting them.

CHAPTER NINE

GROWING BEYOND THE FIRST SEASON

Your first year as a homesteader can be overwhelming. There's a garden to start, a chicken coop to build, and fruit trees to plant. Although these are all great plans, you don't have to take on every project in your first season.

Perhaps your focus this year is a vegetable and herb garden. That's a fantastic place to start because you can grow a lot of food for your investment. Make the most of it by extending the growing season and preserving the harvest for winter. This chapter covers how to set goals and to plan and explains which tools you need to raise more food and prepare the harvest for your pantry.

An important part of learning and planning is keeping track of vegetables that produce well, techniques that succeed or fail, and backyard areas that are best suited to your growing needs. Every year reveals a bit more about your microclimates, soil structure, and seasonal conditions. Record your observations, track your harvest totals, and jot down ideas to maximize future production. All this information will help you improve and expand your homesteading skills.

If you want to sell your home-raised harvests, the tips in this chapter will be even more important for tracking your costs so you know how to price your goods. We also review how to find out what you are legally allowed to sell and how to take your products to market.

EXTENDING AND PRESERVING YOUR HARVESTS

Gardening is a great way to raise fresh food for summer, and one or two cold frames or floating row covers can extend your season. Try planting crops in succession, as discussed in chapter 3, so you can harvest vegetables even longer. Save cash by starting your own seedlings under lights indoors in spring. Look for different varieties to extend your season, such as day-neutral strawberries that continue producing after the June bearers are done.

Over the winter months, you can raise fresh sprouts or microgreens. Industrious homesteaders can create an indoor system for growing fresh food year-round.

Plan for dairy, meat, and egg production over a longer period of time. Your hens will continue laying through winter if you keep a light on in their coop. Stagger breeding times for goats for year-round milk production. Preserve meat, vegetables, and fruits for winter meals.

PLANT PROTECTORS AND SEASON EXTENDERS

Raising vegetables and herbs doesn't have to be just a summertime pursuit. In mild climates, you can plant cool-season crops in the fall and harvest them throughout winter. Kale, collards, mustard greens, and lettuce provide fresh salads from a cold frame. In northern areas, there is a lot more work involved in raising fresh greens for your table in freezing weather. Here are a few options:

Greenhouses. A heated greenhouse allows you to grow four seasons of veggies. If you are handy and have a knack for finding free materials, you can construct a greenhouse without spending a fortune. The disadvantage of a greenhouse is the cost to install and heat it over winter.

Cold frames. These enclosures have transparent tops to let light in. They are cheaper than greenhouses for extending your season, and you can construct them with scraps and repurposed windows. Use them to prepare for and plant cool-weather crops for harvest in spring and into winter.

Cloches. These coverings for individual plants are another useful way to extend your season. You can make one by cutting the bottom off a plastic milk jug. Use a small stake to hold the jug in place.

Floating row covers. Floating row covers work great for light frost protection, especially if you have a whole row of plants. They aren't designed to protect warm-season crops from a heavy frost, however. Try using them for setting out broccoli, cabbage, or other brassicas early. Leave the cover in place to prevent cabbage loopers from laying their eggs on the leaves.

Raised beds. These can extend your growing season because the soil heats up earlier in spring and stays warm later in fall than it does at ground level. Cold air settles in low spots, so plants in raised beds are also less susceptible to a light frost.

DRYING, FREEZING, AND OTHER PRESERVATION METHODS

Preserving your surplus harvests increases your self-reliance throughout the year. Jars of preserves, a freezer full of meat and vegetables, dried fruits and herbs, and roots in cold storage provide food all winter. If you are new to home food preservation, it might sound complicated and difficult. But once you've mastered the art of canning, dehydrating, and pickling foods, the process will be second nature.

The following table lists common preservation techniques, which foods work best for each, storage length, and pros and cons. Use this to help you decide which methods best suit your needs, then read up on how to get started.

COMMON HOME PRESERVATION TECHNIQUES

PRESERVATION METHOD	WHICH FOODS?	STORAGE LENGTH	PROS	CONS
WATER-BATH CANNING	High-acid foods, such as fruits and pickles	1 to 2 years	Shelf-stable; faster than pressure canning	Not safe for low-acid foods
PRESSURE CANNING	Low-acid foods such as meats and vegetables	1 to 2 years	Shelf-stable	The needed equipment can be expensive
FREEZING	Fruits, vegetables, herbs, meats, eggs, dairy, prepared foods, baked goods	1 month to 1 year	Easier and faster than canning; retains more vitamins; fresher flavor	The freezer and the electricity to run it can be expensive; not shelf-stable
DEHYDRATING	Fruits, vegetables, herbs, meat (jerky)	1 year	Easier than canning; shelf-stable	Food can get moldy or rancid if it is not dried properly
FERMENTING	Fruits, vegetables, wine, beer, fermented dairy products	1 week for dairy products to 1+ year for wine	Increases natural probiotics in food; potential for home-brewed wine and beer	Some products are not shelf-stable and there is potential for improper bacteria or yeast culture to ruin flavor

ROOT CELLAR	Root crops, apples and pears, squash	1 to 6 months	No processing or electricity needed; store foods that last naturally in cold storage	Need a cold storage area that doesn't freeze; check regularly for rotting food

How to Freeze Food

Freezing food is one of the easiest methods of preservation. You can store meat, dairy, eggs, fruit, vegetables, herbs, and even baked goods, casseroles, and other foods for one month to a year without losing much nutritional content. Before freezing, you must blanch vegetables in hot water or steam for one to three minutes to destroy enzymes that cause food to deteriorate. Pack the food into BPA-free freezer bags or containers and label them with the date and contents. For the best flavor, use baked goods, eggs, dairy, and prepared foods within one to two months. You can freeze most other foods for up to one year. Prepare fruits by peeling, coring, pitting, and slicing or chopping. Mix in a small amount of lemon juice and sugar and pack in airtight containers. Milk can be frozen as long as you use plastic containers and allow space for it to expand. Butter keeps nicely in the freezer. To freeze eggs, beat first and add ½ teaspoon of sugar or a pinch of salt for every two eggs to preserve their texture for baking or cooking.

How to Dehydrate Food

Fruits, vegetables, and herbs are easy to dry for shelf-stable storage. You can also dry meats as jerky, but use a preservative to prevent spoilage. You can slice most fruits and place them in your dehydrator without a preservative or blanching treatment, but you may want to dip them in lemon juice to prevent browning. Place thin slices of fruit on the trays and dry at a low temperature until dry but still pliable. Before drying cherries and grapes, slice them in half and remove pit or seeds. Vegetables should be blanched for one to three minutes to destroy enzymes. Herbs are one of the easiest foods to dry. Strip leaves and place in the dehydrator on low, or hang small bunches in a cool, dark, dry place.

How to Ferment Food

Wine, beer, kombucha, and yogurt are fermented foods you can make at home. Fermented products use a special strain of yeast (beer, wine, sourdough starter for bread), bacteria (yogurt, sour cream, cheese), or, in the case of kombucha, both yeast and bacteria. Food will ferment with wild yeast and bacteria when left uncovered, but this produces random results. Yogurt is an easy fermenting project to start with; see here for instructions on making your own.

You can produce some fermented foods and drinks by mixing with a culture, placing in a sterile jar, and capping with an airlock to allow the gas to escape and prevent contamination with wild yeast or bacteria.

How to Store Food in a Root Cellar

A root cellar is typically a room in the basement where the winter temperature is kept just a few degrees above freezing. Not many people have root cellars these days, and adding one can be expensive. If you are handy, however, you may be able to build one in a corner of your basement by framing in a room, insulating it, and adding a source of airflow to the outdoors. One vent is drilled through the wall to allow cold air into the root cellar, and a second vent allows warm air out. The best storage temperature for most roots is between 33°F and 40°F with fairly high humidity, so you will need to insulate the ceiling of this space to prevent chilling the rest of the house.

Store fruits and vegetables separately in shallow crates or boxes that allow airflow. Carrots, beets, and turnips do best layered in damp sand to prevent moisture loss. Set pumpkins and winter squash on cardboard or shelves rather than cement floors, where they will rot. Garlic and onions prefer a warmer, dryer storage spot and should do well stored in mesh bags at normal basement temperatures for most of winter.

PRESSURE CANNING AND WATER-BATH CANNING

Safety concerns are probably the biggest reason people avoid canning their harvests, which is a very efficient form of home food preservation. Following these practices will keep you safe while canning and consuming your preserves. The recipes in this book are based on methods updated as of the time of writing, but always follow current guidelines. A good source of safe information is the National Center for Home Food Preservation website, which recommends these best practices:

- Always use updated canning techniques and recipes.

- Find out your elevation and follow proper processing times and pressures.

- Check your equipment yearly for safety.

- Use only jars and lids made specifically for canning.

- Check all canning jars for cracks or chips.

- Use canning lids only once; screw bands may be reused.

- Use a pressure canner with the Underwriter's Laboratory (UL) approval.

- Check dial gauges each year for accuracy.

- Use a pressure canner with a pressure release valve.

- Make sure the pressure has returned to normal before opening the pressure canner.

- Use only fresh foods.

Knowing your elevation, or how many feet you are above sea level, is essential for determining an accurate processing time for water-bath canning and the pressure per square inch (psi) to use for pressure canning in your area. Most recipes assume an elevation of 0 to 1,000 feet for the processing time or psi. If you live at a higher elevation, you need to increase your processing time or the psi to can foods safely. Consult the following table for more information.

THE EFFECTS OF ELEVATION ON PROCESSING TIMES AND PSI

ELEVATION ABOVE SEA LEVEL IN FT.	WATER-BATH ADJUSTMENTS: INCREASE PROCESSING TIME BY . . .	PRESSURE CANNER ADJUSTMENTS: WEIGHTED GAUGE	PRESSURE CANNER ADJUSTMENTS: DIAL GAUGE
0 to 1,000	0 minutes	10 psi	11 psi
1,001 to 2,000	5 minutes	15 psi	11 psi
2,001 to 3,000	5 minutes	15 psi	12 psi

3,001 to 4,000	10 minutes	15 psi	12 psi
4,001 to 6,000	10 minutes	15 psi	13 psi
6,001 to 8,000	15 minutes	15 psi	14 psi
8,001 to 10,000	20 minutes	15 psi	15 psi

If you aren't sure whether older canning equipment is safe, check with your local Cooperative Extension Service office, which can check dial gauges on pressure canners for safety and accuracy. Here are the tools to have on hand for water-bath canning and pressure canning:

♦ Water-bath canner with rack

♦ Pressure canner with rack and functioning gauge, pressure release valve, and seal

♦ Jar lifter

♦ Jar funnel

♦ Jars made for home canning

♦ Single-use canning lids and screw bands

WATER-BATH CANNING VS. PRESSURE CANNING

High-acid foods are safe to preserve in a water-bath canner, but low-acid foods must be pressure canned for safe storage. Here are lists of common high- and low-acid foods:

High-Acid Foods Safe for Water-Bath Canning

↗ PICKLES AND RELISH MADE WITH VINEGAR

↗ JAMS, JELLIES, MARMALADE, AND FRUIT PRESERVES

↗ MOST FRUITS

Low-Acid Foods Safe for Pressure Canning

↗ MEAT, POULTRY, AND SEAFOOD

↗ VEGETABLES

↗ TOMATOES WITHOUT ACID ADDED

↗ SOUPS MADE WITH MEAT AND VEGETABLES

GENERAL CANNING INSTRUCTIONS

When it comes to water-bath and pressure canning, there is some overlap. Follow these initial instructions for both:

1. Get your ingredients ready and read the recipe.

2. Wash canning jars and screw bands thoroughly.

3. Check the manufacturer's instructions for preparing canning lids.

4. Raw pack: Slice or chop food and pack loosely into jars and add hot water up to the recommended headspace.

5. Hot pack: Prepare food and bring to a boil, then ladle into clean, hot jars.

6. Use a knife to remove air bubbles and check headspace.

7.   Wipe the rim of each jar with a clean cloth.

8.   Place a new lid on the jar and screw on the canning band.

9.   After removing jars from the canner, do not press down on lids or tighten bands until they are cool.

10. After 12 to 24 hours, check the lids to make sure they sealed. Press down on the lid. If it gives (moves up and down), the jar is not properly sealed and you should refrigerate the food or use it immediately.

11. Remove the screw band and wash the jar to remove any food on the outside of it.

12. Store canned foods in a cool, dry, dark place.

WATER-BATH CANNING HIGH-ACID FOODS

If you are working with a high-acid food that requires water-bath canning, follow these steps:

1.   Follow the general canning instructions.

2.   If the processing time is less than 10 minutes, sterilize the jars in boiling water for 10 minutes.

3.   Place the rack in the canner.

4.   Heat the water in the canner to 140°F for raw-packed foods or 180°F for hot-packed foods.

5.   Prepare your food according to the recipe.

6. Use the jar lifter to place the filled jars in the canner.

7. Make sure the jars are covered by at least 1 inch of water; add boiling water if needed.

8. Place the lid tightly on the canner.

9. When the water reaches a full rolling boil, begin processing time.

10. If the water stops boiling at any point, return it to a boil and restart processing time.

11. When processing time is up, use the jar lifter to remove the jars, being careful not to tip them.

12. Set jars 2 or 3 inches apart on a towel to cool.

PRESSURE CANNING LOW-ACID FOODS

If you are working with a low-acid food that requires pressure canning, follow these steps:

1. Follow the general canning instructions.

2. Follow the pressure canner manufacturer's instructions for safe use.

3. Add water to the bottom of the canner according to the manufacturer's instructions.

4. Place the rack in the bottom of the canner.

5. Place the filled jars in the canner.

6. Place the lid on the canner and line up the top and bottom so it will seal properly.

7. Turn on the heat under the canner.

8. Allow steam to vent from the pressure valve for 10 minutes.

9. Place the gauge on the pressure valve.

10. Begin the processing time once the pressure has reached the proper level. Maintain pressure throughout the processing time.

11. If pressure drops lower than the required level, restart processing time.

12. When the processing time is complete, turn down the heat and allow the pressure to return to normal.

13. Do not remove the pressure gauge or lid until the pressure has returned to normal.

Troubleshooting Tips: If food leaks out of the jars during pressure canning, the pressure inside the canner was reduced too quickly. Increase the temperature slowly and maintain a steady heat. Do not turn the heat up and down during processing. When the processing time is over, reduce the heat gradually to prevent uneven pressure.

MONITORING AND REASSESSING

Projects can help you gain experience, become self-reliant, and make money. Increase your chances of success by monitoring each project, tracking your progress, and reevaluating your methods.

You won't know if you are saving money on a project if you don't monitor expenses and track outcomes. Weigh your harvests and record them. At the end of the year, add up the cost of your supplies and your total yields to find the cost per pound of your harvest. Did you save money over buying the same foods from the store? If not, what went wrong? Your notes about late spring frosts, heavy rainfall, pests and disease, and other problems will likely offer valuable insights.

Track each project on your homestead to determine which meet your goals and which need adjusting. Projects that don't produce a return on your investment may not be worthwhile. If you've set a goal to raise farm-fresh eggs for the farmers' market, add up the costs to buy hens and feed and follow the regulations in your area for farm-fresh egg sales. Detailed records and regular evaluation of expenses, goals, and methods allows you to learn from experience.

Some of the most useful and inexpensive tools for homesteaders are a computer, a notebook, and a calendar. Use them to record the following information:

- GOALS AND THE STEPS TO ACHIEVE THEM
- COST ESTIMATES FOR EACH PROJECT
- TIME FRAMES FOR COMPLETING TASKS
- DAILY, MONTHLY, AND SEASONAL CHORES
- EXPENSES
- YIELDS
- PROBLEMS ENCOUNTERED, SOLUTIONS TRIED, AND OUTCOMES
- IDEAS FOR INCREASING YIELDS AND REDUCING COSTS
- CROP AND PASTURE LOCATIONS AND ROTATION STRATEGIES
- STEP-BY-STEP INSTRUCTIONS FOR NEW PROJECTS
- REGULATIONS FOR SELLING PRODUCTS OR PROVIDING SERVICES

PURSUING OTHER HARVEST OPPORTUNITIES

The projects discussed in this book are a great place to start when planning your homestead goals and projects, but there are many other ways to save money, increase self-sufficiency, and earn a side income. Here are a few ideas for ways to grow more for your family or to increase your income from the homestead:

- Start seedlings for your garden or for sale.

- Raise cut flowers for sale.

- Provide farm-sitting services.

- Forage for wild herbs, greens, fruits, and roots.

- Hunt and fish to supplement your harvests.

- Create crafts or bath and body products for sale.

- Start an aquaculture system for fish and vegetables.

For every project you want to try, it's important to determine feasibility, create specific goals, and plan your steps. In addition, look into whether you can scale up a small project for a larger growing space or business venture. Will the local population support the increased availability? Or will you saturate the market, reducing the value of your product?

RUNNING A SIDE HOMESTEAD BUSINESS

Starting a homestead-based business can be a satisfying way to increase your income and provide locally raised foods for your community. You may want to set up a farm stand or sell at a farmers' market. This is a wonderful goal, but it takes a lot of planning, work, and outreach to make a small business successful.

Get off to a good start by researching your potential customers, your competition, the prices your market will support, and regulations regarding your future business. A business plan is crucial for focusing your energies and applying for a small business loan. You may also need a license, inspections, and insurance to meet legal requirements and protect your livelihood.

See if your local small business development office provides free consultations or classes in starting your own business. These can be great resources for budding entrepreneurs.

HOW TO IDENTIFY YOUR BUYERS

When you're planning a business venture, it's vital to know who will purchase your products or use your services. Your clientele and their purchasing decisions will influence all of your business decisions. Answer the following questions to get a better idea of who will purchase your products and what their spending habits are:

- Who needs your product or service?

- Where do they shop for similar products?

- Where do they live?

- Do they have disposable income?

- What forms of media do they consume?

If you are struggling to answer these questions, do additional research to narrow down your target customer. Create a survey, join local forums or online groups, ask around, or scout out your competition's customers to fill out the profile of your target audience. Gather the largest sample of answers possible. Determine your secondary target audience, too.

For example, say you want to learn more about customers who attend your local farmers' market because you hope to sell eggs and vegetables there. Visit the market over several months and take notes. Stay for the duration of the market and pay attention to who is shopping, what they purchase, how much they spend, and when the stalls are busiest. Note age range, gender, and any other details that seem important.

BEST PRACTICES FOR SELLING YOUR PRODUCTS

There's an art to providing a product or service from your homestead and making your business profitable. To be successful, you need to be part farmer and part salesperson. Not only must you raise a product at a low price, but you also have to market your goods and develop a good relationship with your customers.

Print business cards and create a website with inviting photos and news from your homestead. If you have a farm stand, keep it stocked with fresh goods. Learn as much as you can about your business so you can speak with authority. Consider giving talks at local business meetings or offering classes at a community college. Ask your friends and family to spread the word, and look into advertising and signage as part of your marketing.

Selling products and offering services in a local market has pros and cons. You can interact with your clients and get to know what they purchase, what they are searching for, and where they like to shop. In turn, they can see the quality of your products firsthand and feel confident you're supplying exceptional products at a reasonable price. The downside is that your customer base may be limited or have low income.

If you plan to sell at a farmers' market, check out each one in your area and note which have the best turnout and sales.

Here are some great ways to make your booth inviting:

♦ Use a pop-up canopy for shade and protection from rain.

♦ Set up folding tables and cover them with red-and-white checkered tablecloths.

♦ Use clean, attractive baskets to display goods.

♦ Sell cut flowers to attract customers.

♦ Display signage that clearly marks your wares and prices.

♦ Hang a banner on your canopy with your farm name in bold letters.

♦ Use a chalkboard to list products for sale, what's coming next week, or specials.

♦ Print the name of your farm and website on bags or a pamphlet for marketing.

♦ Strive to offer a variety of products each week, especially in spring and fall, when other booths may be limited.

♦ Print recipes that feature your weekly specials.

- ◆ Encourage people to visit your website and sign up for news from your homestead.

- ◆ Remember that nothing sells like the color red, abundance, and a friendly demeanor!

UNDERSTANDING YOUR HOMESTEAD MARKET

Get to know the other farmers in your area and familiarize yourself with common agricultural terms. Offering a high-demand product can drive sales, but if surrounding farms have the same thing, you may not get a good price. Use the following labels only if you can demonstrate that your products are certified or adhere to the current restrictions listed on the USDA website. Improper labeling can result in fines.

Certified organic. Products have been raised according to federal organic certification guidelines, including using certified organic seeds, agricultural methods, weed control, irrigation water, and animal feed, among other things. Farmers must keep meticulous records of their process and submit to on-site inspections on a regular basis. Certification requires application fees, inspection fees, and annual certification fees.

Natural. No artificial ingredients or colors are added, and the product has had very little processing. Products must be labeled with a statement explaining the meaning of the term "natural."

Free-range. Free-range poultry must be allowed access to the outdoors.

Grass fed. Small farms may label meat as grass fed if cattle, goats, or sheep have eaten only grass and forage. Animals may not be fed grain and must have access to pasture.

No antibiotics. Meat and poultry may be labeled with "no antibiotics added" if the producer has documented and can show they have not administered antibiotics to the animals.

No hormones. Beef may be labeled "no hormones administered" if the producer has documented and can show the animals have not received hormones. Poultry and pork may not be produced with hormones and may only be labeled with "no hormones" if the label states that hormones are not allowed by law in the production of poultry and pork.

Kosher. Meat and poultry may be labeled kosher only if they have been processed under the supervision of a rabbi.

DETERMINING YOUR PRODUCT PRICES

You'll need to keep detailed records of the amount of time you spend raising, harvesting, and packaging your products to determine your compensation for labor. Also keep detailed records of your investment in seeds, fertilizers, packaging, marketing, and farmers' market fees. Many products you raise may have a short shelf life, so add in your losses.

Use the following formula to determine the price of your products: *(hours of labor × hourly wage + cost of materials and marketing) / yield = sale price.*

It can be tricky to determine some of these figures, and once you calculate the price of your product, you need to compare it to the going rate in your area. Don't undersell other farmers or price yourself out of the market. Livestock auctions, breeders, classified ads in the local paper, and Craigslist listings in your area are a good place to start when selling farm animals.

LEGAL REQUIREMENTS

This is not intended as legal advice. Please contact a lawyer or other licensed legal counsel for legal advice pertaining to your homestead business.

Legal requirements are often the sticking point for a new homestead business. You must adhere to federal, state, and local regulations in all aspects. Some business models have just a few legal requirements, but others are strictly regulated. Start by calling your county health department for restrictions on sale of meat, eggs, dairy, produce, and other products. Here are some basic steps you can take:

- Visit the USDA National Agricultural Library Laws and Regulation website for federal regulations.

- Visit your state and county websites for regulations.

- Call your county health department for rules regarding sale of food products.

- Check whether you need any inspections for fire codes, health department codes, and building codes before you begin operating a business.

- Check into EPA regulations regarding disposal of manure and use of pesticides.

- Check the status of cottage food laws in your area to determine whether you can produce and sell low-risk foods.

- Find out if you have to collect, report, and pay local and state sales taxes.

- Write a business plan.

- Determine whether you should operate as a sole proprietor, limited liability company, or corporation.

- Choose a business name and register with your state or county clerk, or file your paperwork with your state.

- File for appropriate licenses and permits.

Regulations may seem intrusive or difficult to meet and laws can be confusing, but they are created to protect the public from misleading claims or harmful business practices. You can be fined for noncompliance even if you are not aware of a law.

Your reasons for homesteading will be the ultimate guide for the projects you choose and the amount of time you spend raising your own food. To avoid burnout, start out small and work your way up to a self-sufficient lifestyle or a homestead side business. Don't worry if you can't do everything right away. The journey is half the fun.

APPENDIX

VEGETABLE AND HERB RECIPES

Now that you're raising fresh, delicious vegetables and herbs in your backyard, the next step is preparing the harvest for your table. Learn to cook from scratch using basic ingredients and preserve the extra produce for winter to save money, reduce waste, and live more healthfully. Here are recipes using your fresh vegetables and herbs for a variety of dishes, from easy pickles and a winter vegetable salad to canning your own spaghetti sauce and making a scrumptious dessert.

Colorful Cabbage and Root Salad

SERVES 5 / PREP TIME: 25 MINUTES, PLUS 1 HOUR TO CHILL

Make a healthy and hearty salad with vegetables from your garden. This dish is a great way to add nutrients to your diet and increase your self-reliance. Why rely on the grocery store to supply your winter produce when you can grow and store your own? You'll need a grater to shred vegetables and a nonreactive bowl for your salad.

2 cups shredded red or green cabbage

1 cup shredded golden beets

½ cup shredded carrots

¼ cup finely chopped onion

1 garlic clove, crushed

½ cup chopped hazelnuts or almonds

¼ cup red wine vinegar

¼ cup extra-virgin olive oil

1 tablespoon toasted sesame oil

3 tablespoons honey

Salt

Freshly ground black pepper

1. Put the cabbage, beets, carrots, onion, and garlic in a large nonreactive bowl. Add the nuts.

2. In a small bowl, whisk together the vinegar, olive oil, sesame oil, and honey.

3. Pour the dressing over the salad and toss to combine. Season with salt and pepper to taste. Serve right away or refrigerate for 1 hour or overnight to allow the flavors to blend.

Tip: Try adding adobo seasoning for a different flavor.

Brine-Cured Dill Pickles

MAKES 32 TO 40 PICKLES / PREP TIME: 30 MINUTES, PLUS 2 TO 3 DAYS TO BRINE

Make an easy batch of pickled vegetables the same way our grandparents did. The brine solution prevents spoilage while your cucumbers or other vegetables cure. These pickles make a delicious and easy condiment for topping sandwiches or preparing relish. You'll need a nonreactive bowl or crock for brining. Check your local cottage food laws to determine whether the sale of prepared vegetables is allowed in your area.

2 cups water

2 cups distilled white vinegar

¼ cup nonionized coarse sea salt

1 tablespoon whole black peppercorns

2 or 3 dill flowers, washed well

8 to 10 cucumbers or other fresh vegetables, cut into ½-inch slices

1. In a medium saucepan, warm the water, vinegar, and salt over medium heat. Stir until the salt has dissolved.

2. Put the peppercorns, dill flowers, and cucumbers in a large glass or stainless steel bowl or in a pickle crock. Add the brine solution, making sure it completely covers the vegetables.

3. Cover the cucumbers with a nonreactive plate (glass or plastic works well) that just fits inside the bowl and holds them under the brine. If they come into contact with air, they are susceptible to mold and rot.

4. Brine the cucumbers for 2 or 3 days at room temperature. Use a fork to remove one, then taste it. If you are happy with the flavor, transfer the pickles to clean canning jars and cover with brine. Refrigerate and use within a month for best flavor.

Tip: If the pickles have an unpleasant odor or feel slimy, toss them in the compost. Iodized salt gives the pickles an off color, so be sure to use salt without iodine.

Mediterranean Veggies with Pasta

SERVES 10 / PREP TIME: 30 MINUTES / COOK TIME: 30 MINUTES

This dish is easy to make in summer when you're busy with the harvest. It's great at the height of the season, when your eggplant comes in, but you can use whatever is ripe. You'll need a wok or large skillet for the vegetables.

2 quarts water

Salt

3 cups (8 ounces) rotini pasta

3 to 4 tablespoons extra-virgin olive oil, plus more for the pasta water

1 cup chopped onion

1 garlic clove, minced

2 cups peeled and chopped eggplant

2 cups seeded and chopped bell peppers

2 cups chopped tomatoes

1 cup chopped fresh basil, plus more for garnish

Freshly ground black pepper

Shredded cheese, for topping

1. In a large pot over high heat, combine the water, a drop of olive oil, and a pinch of salt. Cover and bring to a boil. Add the pasta, stir, and bring back to a boil. Reduce the heat to medium-high and cook, uncovered, for 7 minutes or according to package directions, until al dente. When the pasta is done, drain it in a colander.

2. In a large wok or skillet, heat the olive oil over medium-low heat until it shimmers. Add the onion and garlic and cook for 10 to 15 minutes, until

the onion is translucent. Turn the heat up to medium.

3.   Add the eggplant and bell peppers and cook for 15 minutes, until almost tender. Add the tomatoes and basil and cook, stirring occasionally, for 10 minutes, until bubbly. Season with pepper to taste. Reduce the heat to low and stir until the desired consistency is reached.

4.   Mix the pasta and veggies together, or place them in separate serving bowls and dish up at the table. Top with the shredded cheese and basil, if desired.

Tip: Add extra garlic and onions or fresh oregano, thyme, and summer savory from the garden. You can use other pastas, such as linguine, shells, or bow tie, if you prefer.

Spaghetti Sauce with Canning Instructions

MAKES 9 TO 10 PINTS / PREP TIME: 1 HOUR / COOK TIME: 3+ HOURS, PLUS 25 MINUTES TO PROCESS

Make your own sauce with ripe tomatoes and fresh garlic, onions, and peppers from the garden. Process homemade sauce in your pressure canner to preserve the bounty of your garden. Check your local cottage food laws to determine whether the sale of low-acid canned goods is allowed in your area.

30 to 40 pounds paste tomatoes, chopped

3 tablespoons extra-virgin olive oil

5 green bell peppers, chopped

3 garlic cloves, crushed

1 cup finely chopped onions

¼ cup sugar

1 tablespoon Italian seasoning

Salt

1. In a large stainless steel pot, cook the tomatoes over medium-high heat for 20 minutes, or until tender. Transfer to a large bowl to cool.
2. Put the olive oil, bell peppers, garlic, and onions in the pot over medium-high heat and cook for 15 minutes, until tender. Remove from the heat.
3. Run the tomatoes through a food mill.
4. Transfer the tomatoes back to the pot. Add the sugar and Italian seasoning, season with salt to taste, and bring to a simmer over medium-

low heat, stirring occasionally. Reduce the heat to low, then place a heat diffuser under the pot to prevent scorching. Cook until the sauce is thick or until the volume is reduced by half. This can take most of the day, depending on the water content of the tomatoes.

5. Fill clean pint-size canning jars with the hot sauce, leaving 1 inch of head space. Wipe the rims, adjust the lids, and process the jars for 20 minutes at 10 pounds of pressure on your pressure canner gauge (0 to 1,000 feet of elevation) or 15 pounds of pressure (1,001+ feet of elevation). Process quart jars for 25 minutes at 10 pounds of pressure (0 to 1,000 feet elevation) or 15 pounds of pressure (1,001+). Follow your pressure canner directions.

Tip: If you want to add meat to your spaghetti sauce, you will need to brown the meat before adding it and increase the processing time to 90 minutes at 10 pounds of pressure (0 to 1,000 feet of elevation) or 90 minutes at 15 pounds of pressure (1,001+ feet of elevation).

Carrot-Zucchini Double Layer Cake

SERVES 12 / PREP TIME: 30 MINUTES / COOK TIME: 35 MINUTES

This moist cake is a delicious way to use your homegrown carrots and zucchini. Make a double layer round cake and top with cream cheese frosting for a special occasion. You won't have any trouble getting your family to eat their veggies with this recipe. Check your local cottage food laws to determine whether the sale of baked goods is allowed in your area.

2 cups all-purpose flour, plus more for pan

1 teaspoon baking powder

1 teaspoon baking soda

1 teaspoon ground cinnamon

¼ teaspoon salt

¼ teaspoon ground nutmeg

1½ cups grated carrots

1½ cups grated zucchini

1¾ cups sugar

1 cup vegetable oil

4 medium eggs

6 ounces cream cheese, at room temperature

½ cup unsalted butter, at room temperature, plus more for greasing

2 teaspoons pure vanilla extract

4½ cups confectioners' sugar

1. Preheat the oven to 350°F. Grease and flour a 9-by-13-inch cake pan or two 9-inch round cake pans and set aside.
2. In a large bowl, sift together the flour, baking powder, baking soda, cinnamon, salt, and nutmeg and set aside.
3. In a medium bowl, combine the carrots, zucchini, sugar, oil, and eggs.
4. Add the vegetable mixture to the flour mixture and stir until thoroughly combined.
5. Pour the batter into the prepared pan and bake for 30 to 35 minutes, or until a toothpick inserted near the center comes out clean.
6. Let cool on a wire rack for 10 minutes, then remove from the pan to cool completely.
7. In a medium bowl, use an electric mixer to beat together the cream cheese, butter, and vanilla until fluffy. Add the confectioners' sugar in small amounts, beating with the electric mixer to combine after each addition until the frosting is spreadable.
8. When the cake has cooled completely, spread the cream cheese frosting over it evenly. Cover and refrigerate until ready to serve.

Tip: Freeze unfrosted cake for up to 1 month for an easy dessert.

FRUIT RECIPES

What could be better than plucking a fresh apple from the tree and eating it right out of your hand? Having your own orchard, small fruits, and nut trees gives you the opportunity to enjoy fresh fruit, make pies and cobblers, create jams and jellies, shell nuts for eating and cooking, and preserve enough fruit for the winter. Try some delicious recipes from purchased fruits and nuts while you wait for the bounty of your own homestead to come in. You'll get some experience with cooking from scratch and preserving fruit from a u-pick, stand, or farmers' market.

Crème de Menthe Citrus Salad

SERVES 8 / PREP TIME: 30 MINUTES, PLUS 1 HOUR TO CHILL

This fresh fruit salad made from oranges and grapefruits with a splash of crème de menthe is a wonderful side dish for a special brunch or any other meal. Try this in winter when other fresh fruits are in short supply. Garnish with a sprig of mint to dress up this wonderful salad for company.

4 oranges

2 pink grapefruits

1 (16-ounce) can pineapple chunks, drained

1 shot glass crème de menthe

Maraschino cherries, for garnish

Mint leaves (optional)

1. Peel and section the oranges and grapefruits and remove the membranes. Put the fruit in a serving bowl. Add the pineapple.
2. Add the crème de menthe and stir gently to combine.
3. Garnish with a few maraschino cherries and mint leaves (if using). Chill for 1 hour before serving.

Tip: If you don't have crème de menthe, substitute ½ teaspoon of mint extract and 1 tablespoon of sugar. Serve as a topping for pancakes or French toast.

Blueberry-Raspberry Stuffed Crepes

SERVES 9 / PREP TIME: 20 MINUTES / COOK TIME: 35 MINUTES

Crepes sound like a fussy dish to make, but they are very easy and you can stuff them with a variety of savory or sweet fillings. Serve them for breakfast with your own fresh or frozen fruit. Try cherry pie filling, fresh strawberries, or peaches in place of the blueberries and raspberries.

2 cups blueberries

2 cups red or black raspberries

1 tablespoon freshly squeezed lemon juice

½ cup sugar

3 tablespoons cornstarch

1½ cups whole milk

1 cup pastry flour or all-purpose flour

2 medium eggs

1 tablespoon vegetable oil, plus more for greasing

Pinch salt

1. In a medium saucepan, heat the blueberries, raspberries, and lemon juice over medium-high heat. Add just enough water to cover the bottom of the pan to prevent scorching. Cook, stirring occasionally, for about 10 minutes, until the mixture begins bubbling.

2. Transfer ½ cup of the mixture to a small heatproof bowl. Add the cornstarch and combine thoroughly to prevent lumps. Slowly pour the cornstarch mixture into the pan and stir until the fruit filling comes to a boil and thickens. Remove from the heat and set aside.

3. In a medium bowl, combine the milk, flour, eggs, vegetable oil, and salt and beat with a hand mixer just until the batter is smooth.

4. Lightly grease a small skillet with oil or butter and put it over medium heat.

5. When the oil or butter shimmers, pour 2 or 3 tablespoons of batter into the skillet and tilt to coat the bottom with batter. Return the skillet to the burner and cook for 1 to 2 minutes, until the bottom of the crepe is lightly browned.

6. Tilt the skillet to slide the crepe onto the plate. Spoon some fruit filling across the center of the crepe, roll it up, and top with another spoonful of filling.

7. Repeat with the remaining batter, adding oil or butter as needed to coat the bottom of the skillet.

Tip: Use cherries, peaches, oranges, or other fruit for filling crepes. Top with whipped cream; sprinkle with chopped hazelnuts, almonds, or other nuts; and serve for dessert, breakfast, or brunch.

Home-Canned Maple Spiced Applesauce

MAKES 7 TO 8 QUARTS / PREP TIME: 1 HOUR / COOK TIME: 30 MINUTES, PLUS 25 MINUTES TO PROCESS

Canning your homemade sauce is a great way to preserve apples for winter. Use a mixture of apple varieties for a complex flavor and sweeten with maple syrup for a special treat. You'll need a water-bath canner and quart-size canning jars. Check your local cottage food laws to determine whether the sale of home preserves is allowed in your area.

½ bushel apples, peeled and cored

1 to 1½ cups pure maple syrup

3 tablespoons ground cinnamon

2 teaspoons ground nutmeg

½ teaspoon ground cloves

1. Put the apples in a large, heavy-bottomed stainless steel pot. Add enough water to cover the bottom of the pan (about 1 inch deep).

2. Cook over medium heat, stirring often to prevent scorching, for 10 to 20 minutes, until the apples are tender. Remove from the heat and let cool.

3. Run the cooked apples through a food mill or strainer to remove chunks for a smooth sauce.

4. Return the apples to the pot over medium-low heat.

5. Add the maple syrup, cinnamon, nutmeg, and cloves and bring to a low boil for 10 to 15 minutes, stirring occasionally to prevent burning.

6. Pack the applesauce in clean, hot quart-size canning jars, wipe the rims, adjust the lids, and process in a water-bath canner for 20 minutes (1 to

1,000 ft.), 25 minutes (1,001 to 3,000 ft.), 30 minutes (3,001 to 6,000 ft.), or 35 minutes (more than 6,000 ft. of elevation).

Tip: If the applesauce is too thick, it may not reach the proper temperature in the middle of the jar. Add a small amount of water or apple juice to the sauce so it is thin enough to pour, then heat to a low boil before packing in jars. For easy applesauce, freeze instead of canning. For chunky applesauce, coarsely chop the apples before beginning and skip steps 3 and 4.

New England Cranberry Spice Bread

MAKES 1 LOAF / PREP TIME: 30 MINUTES / COOK TIME: 1 HOUR

This traditional quick bread recipe from New England is chock-full of cranberries for a tasty and nutritious snack. Slice and serve for a side at Thanksgiving dinner or as a quick breakfast. You'll need a 5-by-9-inch loaf pan.

2 cups whole-wheat pastry flour, plus more for pan

1 cup sugar

1 teaspoon baking powder

½ teaspoon ground nutmeg

¼ teaspoon ground cloves

Pinch salt

½ cup whole milk

¼ cup freshly squeezed lemon juice

2 tablespoons unsalted butter, melted, plus more for greasing

1 large egg

1 cup cranberries, coarsely chopped

1. Preheat the oven to 325°F. Grease and flour a 5-by-9-inch loaf pan and set it aside.
2. In a large bowl, sift together the flour, sugar, baking powder, nutmeg, cloves, and salt and set aside.
3. In a medium bowl, combine the milk and lemon juice and allow it to curdle for 1 to 2 minutes.

4. Add the milk mixture, melted butter, and egg to the flour mixture and beat by hand until well combined.

5. Fold in the cranberries.

6. Pour the batter into the loaf pan and bake for 60 minutes, or until a toothpick inserted in the center comes out clean.

7. Let it cool in the pan for 10 minutes on a wire rack, then run a knife around the edges of the pan to loosen. Remove the bread from the pan and cool completely on a wire rack.

Tip: For a nutty loaf, add ½ cup of chopped hazelnuts, walnuts, or pecans.

Strawberry Graham Cracker Pie

MAKES 1 PIE / PREP TIME: 30 MINUTES, PLUS 1 HOUR TO CHILL

Fresh pie made with your own berries is a wonderful dessert for spring or summer. Raise a few everbearing or day-neutral berries so you can enjoy this luscious pie all summer long. You'll need a graham cracker crust for this recipe.

8 cups strawberries, hulled, divided

⅓ cup water

⅓ cup freshly squeezed lemon juice

1 cup plus 1 tablespoon sugar, divided

2 tablespoons cornstarch

8- or 9-inch graham cracker piecrust

2 tablespoons crushed graham crackers, for topping

Whipped cream, for topping

1. Put 1 cup of berries, the water, and lemon juice in a blender and purée until smooth. Add enough additional water to make 1½ cups, if needed.

2. In a medium saucepan over medium-high heat, combine 1 cup of sugar and the cornstarch. Add the strawberry purée and cook, stirring occasionally, for 5 to 8 minutes, until the mixture is thick and bubbly. Cook and stir for 2 minutes more.

3. Remove from the heat and cool for 10 minutes without stirring.

4. Spread about one-third of the mixture over the sides and bottom of the piecrust. Add about half of the remaining berries. Spoon half of the remaining cooked mixture over the berries. Add the remaining berries and cover with the remaining cooked mixture. Chill for 1 to 2 hours.

5. Combine the crushed graham crackers and remaining 1 tablespoon of sugar and set aside.

6. Before serving, top the pie with whipped cream and sprinkle with the graham cracker mixture. Garnish with whole strawberries, if desired.

Tip: For a homemade option, you can use a baked single-crust pie shell instead of the graham cracker crust. Add 2 tablespoons of finely chopped nuts to the graham cracker sprinkle for a nutty topping.

GRAIN RECIPES

Raising wholesome grains on the homestead can fill your pantry with flour, cornmeal, rolled oats, and cracked grains for baking and cooking. Winter mornings are cozy and warm when you fill up on hot pancakes or bread fresh from the oven. You'll enjoy preparing your own fresh breads, baked goods, and hearty breakfast cereals with the nutty goodness of freshly prepared grains. Invest in a grain mill to produce the most nutritious flours and meals possible.

Buckwheat-Ginger Pancakes

MAKES 20 TO 24 PANCAKES / PREP TIME: 10 MINUTES / COOK TIME: 50 MINUTES

These hardy hotcakes are a filling and delicious breakfast for hungry homesteaders. The buckwheat lends a nutty taste that complements the light ginger flavor, and it's a great way to add fiber to the first meal of the day. A cast-iron griddle is a fantastic choice for preparing these old-fashioned pancakes.

1 tablespoon oil

1 cup buckwheat flour

1 cup pastry flour

2 tablespoons sugar

4 teaspoons baking powder

1 teaspoon ground ginger

Pinch salt

2 cups milk

2 medium eggs, lightly beaten

¼ cup vegetable oil

1. Oil a cast-iron griddle and set over medium-high heat.
2. In a large bowl, sift together the buckwheat flour, pastry flour, sugar, baking powder, ginger, and salt.
3. Add the milk, eggs, and oil and stir until combined. Scrape the bottom and sides of the bowl to incorporate all of the flour, but don't over beat. There will be a few small lumps. If the batter seems too thick, add more milk.
4. Pour ¼ cup of batter onto the prepared griddle. (A large griddle will allow you to cook 3 pancakes at a time.) Fry until the bottom of the pancake is golden brown, about 3 minutes. Flip the pancake and fry until the other side is golden brown, about 2 minutes.
5. Transfer the pancake to a serving platter and repeat with the remaining batter. Serve.

Tip: For a different flavor, substitute whole-wheat flour or finely ground cornmeal for some of the buckwheat flour.

Maple Spiced Granola

SERVES 16 / PREP TIME: 10 MINUTES / COOK TIME: 1 HOUR

This homemade granola has lots of fiber and nutrients to fill an empty stomach and keep you going all morning. Use pure maple syrup or substitute honey from your own beehives. Check your local cottage food laws to determine whether the sale of baked goods is allowed in your area.

3 tablespoons vegetable oil, plus more for greasing

6 cups rolled oats

2 cups chopped nuts of choice

1 to 1½ cups pure maple syrup

1 teaspoon ground cinnamon

1 teaspoon ground nutmeg

½ teaspoon ground cloves

1. Preheat the oven to 300°F. Grease two 9-by-13-inch baking dishes with vegetable oil.

2. In a large bowl, combine the oats, nuts, maple syrup, and 3 tablespoons of oil, mixing until the oats are well coated.

3. Sprinkle the cinnamon, nutmeg, and cloves over the oat mixture, and stir to combine.

4. Divide the mixture between the prepared pans and bake, stirring every few minutes to prevent burning, until the oats have dried out and browned, 45 to 60 minutes.

5. Let cool completely, then transfer to an airtight container. Use within 2 or 3 weeks for the best flavor.
   Tip: Add 1 cup of shredded coconut for a twist. If you don't have maplesyrup, replace it with honey mixed with 1 to 2 tablespoons of water.

Cinnamon-Pumpkin Muffins

MAKES 36 MUFFINS / PREP TIME: 20 MINUTES / COOK TIME: 30 MINUTES

These scrumptious muffins are a delicious way to use your home-raised pumpkins, eggs, and wheat flour. Grind your flour on the fine setting and use right away for the sweetest flavor. Check your local cottage food laws to determine whether the sale of baked goods is allowed in your area.

1½ cups chopped walnuts or pecans

½ cup sugar

6 tablespoons unbleached all-purpose flour

4 tablespoons unsalted butter, melted

3 teaspoons ground cinnamon, divided

½ teaspoon ground cloves

4 cups whole-wheat pastry flour or unbleached all-purpose flour, divided

2 tablespoons baking powder

½ teaspoon baking soda

Pinch salt

2 cups pumpkin purée

2 cups light brown sugar

1 cup whole milk

5 medium eggs, lightly beaten

½ cup vegetable oil or melted unsalted butter, plus more for greasing

1. Preheat the oven to 350°F. Grease 3 standard 12-cup muffin pans or line with paper muffin cups. Set aside.

2. In a medium bowl, combine the nuts, sugar, all-purpose flour, butter, 1 teaspoon of cinnamon, and cloves. Set aside.

3. In a large bowl, sift together 2 cups of pastry flour, the baking powder, the remaining 2 teaspoons of cinnamon, baking soda, and salt. Add the pumpkin, brown sugar, milk, eggs, and oil. Beat with an electric mixer on low speed until well blended, then beat on high speed for 2 minutes more. Add the remaining 2 cups of flour and stir until well combined.

4. Pour the batter into the prepared muffin pans and top each with a spoonful of nut crumble, dividing it evenly.

5. Bake for about 30 minutes, or until a toothpick inserted in one of the center muffins comes out clean.

6. Let cool on a wire rack for 10 minutes, then remove the muffins from the pans and cool completely.

Tip: Try adding 1½ cups of raisins, chopped nuts, or dried cranberries to the batter.

Light and Savory Wheat Bread

MAKES 2 LOAVES / PREP TIME: 30 MINUTES, PLUS 90 MINUTES TO RISE / COOK TIME: 30 MINUTES

Making your own fresh bread is the quintessential homesteading project. From raising wheat at home to threshing and grinding the grain to kneading the dough and baking fresh loaves, you can't get much more down to earth. Add dried herbs from your garden for savory loaves.

3 to 3½ cups unbleached all-purpose flour, divided, plus more for work surface

2½ teaspoons yeast

1¾ cups whole milk

3 tablespoons unsalted butter

1 tablespoon sugar

1 teaspoon salt

1 tablespoon dried rubbed rosemary, sage, or basil

2 cups whole-wheat bread flour

Vegetable oil, for greasing bowl and pans

1. Grease and flour two loaf pans.
2. In a large bowl, combine 2 cups of all-purpose flour and the yeast.
3. In a small saucepan over low heat, warm the milk, butter, sugar, and salt until the butter is mostly melted and the temperature is between 120°F and 130°F.
4. Add the milk mixture to the flour mixture all at once and beat with an electric mixer on low speed just until combined (or mix by hand). Use a

spoon to mix in the dried herbs, whole-wheat flour, and as much of the remaining all-purpose flour as you can stir into the dough.

5. Turn out dough onto a floured surface and knead in enough flour to make a stiff dough that is smooth and elastic. Use vegetable oil or butter to grease the bowl and put the dough back in the bowl. Cover with a clean dish towel and place in a warm place to rise for about 1 hour, or until doubled in size.

6. Punch down the dough. Turn out onto a lightly floured surface, divide in half, and allow to rest for about 10 minutes.

7. Form each dough half into a log shape and place in the prepared pans. Cover and let rise in a warm spot for about 30 minutes, or until nearly doubled in size.

8. Preheat the oven to 375°F and place pans in oven when hot. Bake until the tops are toasty brown and the loaf sounds hollow when tapped, 25 to 30 minutes.

9. Tip the bread out of the pans and let cool completely on a wire rack before slicing.

Tip: Add grated Parmesan cheese for a tasty twist on this recipe.

Egg and Herb Noodles

SERVES 6 / PREP TIME: 25 MINUTES, PLUS 1 HOUR TO DRY / COOK TIME: 5 MINUTES

Make your own egg noodles with savory herbs for a tasty homemade meal. You don't have to buy a pasta machine; just roll out the dough with a rolling pin and cut into strips. These noodles taste amazing with just a bit of butter and garlic salt, or top them off with sauce made from your own tomatoes.

2 cups durum wheat flour, plus more as needed

3 large eggs

1½ teaspoons water, plus more as needed

1 teaspoon extra-virgin olive oil

Salt

Freshly ground black pepper

Italian seasoning or oregano

1. Place the flour on a clean work surface. Make a well in the center and add the eggs, water, and olive oil. Season with salt, pepper, and Italian seasoning to taste.

2. Use your fingers to combine the ingredients. Add more water or flour as needed. The dough should not be sticky.

3. Knead until the dough is smooth and elastic.

4. Wrap the dough in a clean linen towel and rest at room temperature for 10 minutes.

5. Divide the dough in half. On a floured work surface, roll out each half to a ¼-inch thickness.

6. Roll up the dough jelly roll–style and use a sharp knife to cut it into ½- to 1-inch strips. Hang the noodles to dry for at least 1 hour.

7. To cook the noodles, add a pinch of salt and a few drops of vegetable oil to a large pot of water and bring to a boil. Add the noodles and cook for 3 to 5 minutes, until tender.

Tip: Dry noodles completely for storage. Use a dehydrator on the lowest setting to remove moisture. You can store properly dried egg noodles at room temperature for several months.

EGG AND POULTRY RECIPES

Nothing beats the feeling of serving food prepared entirely from your homestead. Use eggs to make custards, cakes, puddings, and omelets from scratch. Cook your own home-processed chickens, ducks, turkeys, and other poultry for a humane source of protein. You'll never want store-bought eggs again.

Duck Egg Nutmeg French Toast

SERVES 8 TO 10 / PREP TIME: 5 MINUTES / COOK TIME: 30 MINUTES

Duck eggs and whole milk give this French toast a rich, custard flavor. Try using your own whole-wheat bread for an even more self-reliant breakfast. Top the toast with homegrown fruit or honey from your bees.

2 tablespoons oil

6 large duck eggs, lightly beaten

¾ cup whole milk or cream

Nutmeg to taste

8 to 10 whole-wheat bread slices

1. In a cast-iron griddle, heat the oil over medium-high heat until it shimmers.

2. In a wide-bottomed bowl, whisk together the eggs and milk. Two at a time, dunk the bread slices into the egg mixture and put them in the skillet. Sprinkle with a little nutmeg and cook on one side for 2 to 3

minutes, until browned. Flip and cook on the other side for 2 to 3 minutes, until browned.

3. Transfer to a serving platter and serve.

Tip: You can substitute chicken eggs for duck eggs. Use 8 or 9 medium chicken eggs in place of 6 duck eggs.

Duck and Apple Salad Sandwich

SERVES 2 / PREP TIME: 15 MINUTES

This recipe makes enough salad for two hearty sandwiches. Alternatively, you can top a tossed salad or scoop out a tomato and fill it with this sweet and savory alternative to chicken salad.

1 cup chopped cooked duck meat

1 green onion, white and green parts, chopped

½ celery stalk, chopped

¼ cup chopped apple

¼ cup mayonnaise

1 tablespoon mustard

Salt

Freshly ground black pepper

1. In a large bowl, combine the duck meat, green onion, celery, and apple.
2. Add the mayonnaise and mustard. Season with salt and pepper to taste. Stir to combine.
3. Spread duck salad on homemade bread for a delicious sandwich. Refrigerate unused portion and use within 1 to 2 days.

Tip: Replace the duck meat with chicken or the chopped apples with halved grapes, if you like. You can increase this recipe to make as many servings as you need.

Potluck Breakfast Quiche

SERVES 12 / PREP TIME: 40 MINUTES / COOK TIME: 40 MINUTES

This quiche is easier to make than a traditional quiche because you don't have to roll out the crust. It also serves more people. You can add other vegetables from your garden or meat to dress it up, or you can make the basic quiche for an easy and filling meal anytime.

2 cups whole-wheat pastry flour

½ teaspoon baking powder

Salt

½ cup plus 1 tablespoon unsalted butter, at room temperature, divided, plus more (or vegetable oil) for greasing

4 to 6 tablespoons cold water

1 cup chopped green onions, white and green parts

8 medium eggs, lightly beaten

2 cups whole milk or cream

Freshly ground black pepper

1 cup shredded Cheddar cheese

1. Preheat the oven to 350°F. Lightly grease a 9-by-13-inch cake pan or casserole dish.

2. In a medium bowl, combine the flour, baking powder, and a pinch of salt. Cut ½ cup of butter into the flour with a pastry cutter or fork until it is the size of small peas. Add enough water to form a ball.

3. Put the dough in the prepared pan and use your fingers to flatten and press the dough to cover the bottom and about halfway up the sides of

the pan.

4. Bake the crust for 15 to 20 minutes, until the dough in the center is set.

5. In a small skillet, melt the remaining 1 tablespoon of butter over medium heat. Add the green onions and cook, stirring occasionally, for about 5 minutes, until tender. Remove from the heat and set aside.

6. Put the eggs and milk in a medium bowl. Season with salt and pepper, then whisk to combine well.

7. Pour the green onions and the egg mixture onto the baked crust and season with salt and pepper to taste. Top with the cheese.

8. Bake for 40 minutes, or until a knife inserted in the center comes out clean. Cool for 10 minutes and serve.

Tip: Chop and sauté 1 cup of asparagus, broccoli, snow peas, or green beans and add them to your quiche with the green onions to use your homegrown veggies. For a heartier quiche, add cooked sausage or bacon.

Chicken and Onion Enchilada Casserole

SERVES 12 / PREP TIME: 15 MINUTES / COOK TIME: 40 MINUTES

This casserole is easy to prepare and makes a savory meal for lunch or dinner. Use leftover chicken if you have it on hand. You'll need tortillas to line the pan for a simple crust.

1 tablespoon vegetable oil for greasing

8 small corn or flour tortillas

4 cups shredded cooked chicken

1½ cup chopped green onions, white and green parts, divided

2 cups cooked black beans, drained

4 cups shredded Monterey Jack cheese

2 cups mild salsa

1 cup sour cream

1. Preheat the oven to 350°F. Grease the bottom and sides of a 9-by-13-inch casserole dish with the oil. Use the tortillas to line the bottom of the pan.

2. Spread the chicken over the tortillas, then top with 1 cup of green onions and the black beans. Spread the cheese evenly over the top, then top with the salsa.

3. Bake for 35 to 40 minutes, or until the casserole is bubbling.

4. Let cool for 10 minutes, then top with the sour cream and the remaining ½ cup of green onions. Serve.

Tip: You can omit the tortillas if you don't have them and serve with a bowl of tortilla chips instead.

Butternut-Ginger Custard

SERVES 12 / PREP TIME: 15 MINUTES / COOK TIME: 45 MINUTES

This delicious custard tastes just like pumpkin pie but uses butternut squash. It is a great way to use squash from your garden and home-raised eggs. It's wonderful for fall or any other time of year.

1 tablespoon unsalted butter for greasing

4 cups puréed butternut squash

½ cup honey or pure maple syrup

1 teaspoon ground ginger

Pinch salt

6 medium eggs, lightly beaten

1¾ cup heavy cream or whole milk

1 teaspoon pure vanilla extract

1. Preheat the oven to 350°F. Grease a 9-by-13-inch baking dish with the butter.
2. In a medium bowl, combine the squash, honey, ginger, and salt. Add the eggs, cream, and vanilla extract, and stir to combine well.
3. Pour the custard mixture into the prepared dish and bake for 40 to 45 minutes, or until the center of the custard is set.
4. Let cool for 1 hour before serving. Store leftovers in an airtight container in the refrigerator for 3 to 4 days.

Tip: Use canned pumpkin purée if you don't have butternut squash. Top with whipped cream or vanilla ice cream, if desired.

MEAT AND DAIRY RECIPES

Once you have fresh milk and your own humanely raised meat, you'll want to put it all to good use. Make your own dairy products like cheese and yogurt from the extra milk. Pasteurize milk by slowly heating to 165°F for 15 seconds to kill bacteria.

Most people aren't used to eating rabbit, so you'll find a recipe here for home-canned rabbit soup to help you use and preserve this valuable source of meat. You'll need a pressure canner to make up a batch, and there are step-by-step instructions (see here) to guide you.

Farmer's Fresh Goat Cheese

MAKES ABOUT 3 CUPS / PREP TIME: 5 MINUTES / COOK TIME: 15 MINUTES, PLUS 1 HOUR
TO DRAIN

Create your own fresh cheese with just three ingredients. Farmer's cheese isn't aged, so you can make a batch with extra goat milk or even store-bought milk. You'll need a thermometer, some cheesecloth, and a colander for this recipe.

1 gallon goat milk

¼ cup distilled white vinegar

Pinch salt

1. In a large saucepan, warm the goat milk over medium heat, stirring to prevent scorching, to 180°F, then immediately remove from the heat.
2. Gently stir in the vinegar, then let the milk mixture rest at room temperature for 10 to 15 minutes, until it curdles. If the milk does not

separate into curds and whey, add 2 more tablespoons of vinegar and stir. Let sit for 10 minutes more.

3. Use cheesecloth to line the inside of a colander and set it over a large pan. Pour the curds and whey into the colander and allow it to drain for 1 hour.

4. Add the salt to the curds and combine. Refrigerate the cheese and use it within 1 week.

Tip: Add fresh chopped herbs for a savory cheese or 1 or 2 tablespoons of honey for a sweet cheese spread. If you want a harder cheese, wrap the curds in the cheesecloth, set in the colander, and place a weight on top to remove more whey. You'll have a lot of whey left over, so use it in place of milk in baked goods or feed it to the pigs for extra protein.

Creamy Goat Milk Yogurt

MAKES 4 (1-CUP) SERVINGS / PREP TIME: 30 MINUTES, PLUS 12 HOURS TO CULTURE

Use fresh milk from your goats to make your own delicious yogurt. Mix it with homemade jam or homegrown fruit and a drizzle of honey from your bees for a delicious breakfast. Add powdered milk to make it nice and thick, or pour finished yogurt into a cheesecloth-lined colander to drain off some of the whey. You can find powdered goat milk next to the powdered cow milk at the grocery store. You'll need a yogurt maker for this recipe.

1 quart whole goat milk

1 tablespoon plain yogurt

¼ cup powdered goat milk

1. In a medium saucepan, warm the goat milk over medium heat, stirring to prevent scorching, to 180°F. Immediately remove from the heat and let it cool to 110°F to 120°F.
2. Transfer about ¼ cup of milk to a small bowl and stir in the yogurt and powdered goat milk.
3. Pour the mixture back into the pan and stir gently to combine.
4. Pour the milk mixture into a yogurt maker or the containers that fit inside the yogurt maker. Turn on the yogurt maker and allow the milk to culture for 10 to 12 hours, or overnight.
5. When the yogurt is set, turn off the yogurt maker and allow it to cool.
6. Refrigerate and use within 1 to 2 weeks.

Tip: There are several reasons your homemade yogurt may have a runny, slimy, or stringy texture. To avoid these problems, make sure you heat the milk to 180°F to kill unwanted bacteria. Do not add the yogurt culture to milk until it cools to 110 to 120°F. If the milk is too hot, it will kill the beneficial bacteria in the culture. Also, make sure the yogurt you use as a culture is fresh to ensure the bacteria is active.

Rabbit Soup for Pressure Canning

MAKES 7 QUARTS / PREP TIME: 90 MINUTES / PROCESSING TIME: 90 MINUTES

Preserve your home-raised rabbits for an easy meal over winter. Canning is a great way to store shelf-stable meals in your pantry. You'll need a pressure canner and mason jars.

1 whole rabbit, dressed

4 quarts water

1 or 2 bay leaves

7 or 8 carrots, sliced

7 or 8 potatoes, chopped

2 medium onions, chopped

2 garlic cloves, minced

Salt

Freshly ground black pepper

1. Wash 7 quart-size mason jars and set aside.

2. In a large stockpot over medium-high heat, combine the rabbit, water, and bay leaves and bring to a boil. Reduce the heat to low and simmer the rabbit for about 40 minutes, until the meat is no longer pink in the thickest part of the thigh. Remove the rabbit to cool, but keep the pot of broth simmering.

3. Add the carrots, potatoes, onions, and garlic to the broth and increase the heat to medium. Season with salt and pepper to taste, and cook for 15 minutes, until the vegetables are almost tender.

4. Remove the rabbit meat from the bones and divide it evenly among the mason jars.

5. Remove the broth from the heat. Ladle the vegetables into the mason jars. Fill each jar halfway with meat and vegetables, then top with broth, leaving 1 inch of headspace. If you need extra liquid to fill the jars, use hot water.

6. Wipe the rims and adjust the canning lids on each jar.

7. Pour water into the pressure canner, according to the manufacturer's instructions. Process the jars in the pressure canner for 90 minutes at 10 pounds of pressure (0 to 1,000 feet of elevation) or 15 pounds of pressure (1,001 feet of elevation or higher). Processing time begins when proper pressure is reached.

8. Store in a cool (50°F to 70°F), dark spot for up to 1 year.

Tips: You may add herbs from your garden to season this soup. Try savory, sage, rosemary, or thyme.

Sausage Pot Pie

SERVES 8 / PREP TIME: 1 HOUR / COOK TIME: 30 MINUTES

This delicious variation of a pot pie is zesty and filling after a hard day of work. You'll need a single pastry crust to top it off and a casserole dish for baking. Use your own vegetables, wheat, lard, and ground pork for a self-reliant meal.

1 cup pastry flour

Pinch salt

⅓ cup lard or unsalted butter

3 tablespoons cold water

2 tablespoons unsalted butter or vegetable oil

½ cup chopped onion

1 garlic clove, minced

1 cup chopped potato

2 cups ground pork

⅓ cup pastry flour or all-purpose flour

1 teaspoon ground sage

1 teaspoon ground savory

½ teaspoon ground chili powder

Salt

Freshly ground black pepper

1 cup vegetable or chicken broth

1½ cups frozen peas and carrots or mixed vegetables

1. In a medium bowl, combine the pastry flour, salt, and lard. Use a pastry cutter to cut the lard into the flour until it forms pea-size lumps. Add the water, 1 tablespoon at a time, and toss with a fork until the flour is moist. Form a ball with the dough and refrigerate while you prepare the filling.

2. Preheat the oven to 425°F.

3. In a large skillet, melt the butter over medium heat. Add the onion and garlic and cook for 10 to 15 minutes, until the onion is translucent. Add the potato and cook, stirring occasionally, for 10 minutes, until tender. Stir in the ground pork and cook thoroughly for about 15 minutes. Drain the excess fat.

4. Stir in the flour, sage, savory, and chili powder. Season with salt and pepper to taste. Cook and stir for 5 minutes, until the flour is incorporated into the mixture.

5. Add the broth and vegetables and cook, stirring occasionally, for 10 minutes, until bubbly and thick. Pour the filling into a 1½-quart casserole dish.

6. Roll out the dough with a rolling pin and trim the edges to about ½ inch larger than the dish. Turn the edges of the crust under and use your fingers to create a fluted edge. Cut 2 or 3 slits in the top of the crust.

7. Place the casserole dish on a baking sheet to catch spills and bake for 20 to 30 minutes, or until the crust is brown and the juices are bubbling.

8. Let sit for 10 minutes before serving.

Tip: You can replace the ground pork with ground beef, turkey, or chicken, if desired. If you are short on time, skip the piecrust and top with shredded cheese or prepared biscuit dough.

Creamy Chocolate Cooked Pudding

SERVES 6 / PREP TIME: 5 MINUTES / COOK TIME: 20 MINUTES

Homemade pudding from fresh milk and eggs is so much tastier than anything from a box. This pudding is rich and creamy, perfect for making pudding pies, filling cream puffs, or eating as is.

⅔ cup sugar

⅓ cup Dutch-process cocoa powder

3 tablespoons cornstarch

2 cups whole milk

1 cup heavy cream

2 large eggs, beaten

1. In a medium saucepan, combine the sugar, cocoa powder, and cornstarch over medium heat. Gradually whisk in the milk and cream and combine well to prevent lumps.

2. Cook, stirring constantly, for about 10 minutes, until the mixture is thick and bubbly. Cook and stir for 2 minutes more, then remove from the heat.

3. Put the eggs in a small bowl and beat. Gradually whisk 1 cup of the hot mixture into the beaten eggs, then return the egg mixture to the pan. Put the pan over medium heat to cook for about 10 minutes, until nearly bubbly, but do not boil. Reduce the heat to low, then cook and stir the pudding for 2 minutes more.

4. Remove from the heat and let cool. Cover and refrigerate for up to 1 week.

Tip: Make vanilla pudding by omitting the cocoa powder. Serve with sliced strawberries or blueberries from your small fruit plantings as a garnish.

HONEY AND BEESWAX RECIPES

Once you've harvested your first batch of honey, you'll want to slather toast with it, sweeten your oatmeal with it, and drizzle it over pancakes for breakfast. If there's any left, store it in clean jars and cap them with lids that have never been used for pungent foods like pickles or garlic. Mason jars with new canning lids work well for home storage. Keep the beeswax for making your own bath and body products and furniture wax and for numerous other uses around your homestead.

Sell honey and beeswax products at your local farmers' market or from a farm stand as a side business. Be sure to check into local regulations regarding the sale of bee products and label everything properly.

Beeswax and Peppermint Lip Balm

MAKES 8 (1-OUNCE) TINS / PREP TIME: 5 MINUTES / COOK TIME: 10 MINUTES

Making your own lip balm is a great way to use beeswax from your hives and control the ingredients in your bath and body products. This project takes only a few minutes and makes enough lip balm to share. You'll need a double boiler to melt the wax and some small lip balm pots or tubes to pour the molten balm into. This balm might make a great value-added product for a side business. Check the regulations in your area first.

½ cup sweet almond, coconut, or extra-virgin olive oil

½ cup grated beeswax

10 to 15 drops food-grade peppermint essential oil

1. Pour about 2 inches of water in the bottom pan of a double boiler and set it over medium-low heat.

2. Place the oil and grated beeswax in the top pan of the double boiler. Heat until the beeswax melts, about 10 minutes. Stir and remove the pan from the heat.

3. Add the peppermint essential oil and stir. Remove the top pan from the double boiler and wipe the water off the bottom of the pan to avoid mixing water into your lip balm. Be careful, as the pan will be hot.

4. Pour the mixture into lip balm containers and let cool.

Tip: Try using different essential oils, like lemon, lime, and sweet orange.

Beeswax Moisturizing and Healing Salve

MAKES ABOUT 2 (4-OUNCE) TINS / PREP TIME: 5 MINUTES / COOK TIME: 10 MINUTES

Create your own hand salves, foot balms, and other skin care products with home-raised beeswax. You can create different scents and consistencies with just a bit of experimentation. You'll need a double boiler to melt the ingredients, tins for pouring the molten salve into, and some decorative labels to list the ingredients. This salve can be a great vehicle for creating value-added products from your home apiary.

1 cup sweet almond, coconut, olive, or sunflower oil

4 to 5 tablespoons grated beeswax

10 to 15 drops of essential oil of your choice

1. Pour about 2 inches of water in the bottom pan of a double boiler and set over medium-low heat.

2. Place the oil and grated beeswax in the top pan of the double boiler. Heat until the beeswax melts, about 10 minutes. Stir and remove from the heat.

3. Add the essential oil and stir. Remove the top pan from the double boiler and wipe the water off the bottom of the pan to avoid mixing water into your salve. Be careful, as the pan will be hot.

4. Pour the mixture into metal containers and let cool.

Tip: Try lemon-lime, sweet orange, or peppermint essential oils for a nice scent. Increase the oil by 2 or 3 tablespoons for a softer consistency and use 15 drops of essential oil.

Honey-Lemon Cream Cheese Spread

MAKES ABOUT ½ CUP / PREP TIME: 10 MINUTES

This deliciously sweet cream cheese spread tastes wonderful on toast, muffins, or pancakes. If you have citrus trees and your own fresh cream for cheesemaking, you can make this spread entirely from scratch. You'll need an electric hand mixer to combine the ingredients.

½ cup (4-ounce package) cream cheese

2 or 3 tablespoons honey

1 teaspoon grated lemon zest

1 tablespoon freshly squeezed lemon juice

1. In a medium bowl, combine the cream cheese, honey, lemon zest, and lemon juice. With an electric mixer on medium speed, cream ingredients together.
2. Store in an airtight container in the refrigerator for up to 2 weeks.

Tip: Try using orange zest and juice for a different flavor.

Spiced Honey Butter

MAKES ABOUT 1 CUP / PREP TIME: 15 MINUTES

What could be more delicious than your own homemade butter combined with the sweet amber goodness from your bees? This honey butter is better than anything you can buy because it contains no artificial ingredients or preservatives. If the honey and butter separate, just use your electric mixer to recombine them.

1 cup unsalted butter, at room temperature

¼ cup honey

1 teaspoon ground cinnamon

½ teaspoon ground nutmeg

¼ teaspoon ground cloves

1. In a medium bowl, combine the butter, honey, cinnamon, nutmeg, and cloves. Beat with an electric mixer on medium speed until well combined.

2. Store in an airtight container in the refrigerator for 2 or 3 weeks or in the freezer for 2 or 3 months.

Tip: This butter is great on toast, pancakes, and waffles. Make extra to give as gifts, and decorate with custom labels from your homestead.

Spiced Honey Cake

MAKES 12 SERVINGS / PREP TIME: 20 MINUTES / COOK TIME: 55 MINUTES

This moist, tender cake is so sweet that you don't need frosting. The spices give it a bit of extra flavor, or you can leave them out for a more traditional honey cake. If you have a sweet tooth, serve this cake with a drizzle of honey. Check your local cottage food laws to determine whether the sale of baked goods is allowed in your area. If so, set up a stand at a farmers' market and try making individual cakes for quick sales to hungry customers.

1 tablespoon unsalted butter for greasing

4 cups whole-wheat pastry flour

1 teaspoon baking soda

3 teaspoons ground cinnamon

1 teaspoon ground nutmeg

½ teaspoon ground cardamom

Pinch salt

1½ cups unsalted butter, at room temperature

2 cups honey

8 large eggs, at room temperature

½ cup sour milk, at room temperature

1. Preheat the oven to 325°F. Grease a 9-by-13-inch cake pan.
2. In a medium bowl, combine the whole-wheat flour, baking soda, cinnamon, nutmeg, cardamom, and salt and set aside.
3. In a large bowl, beat the butter, honey, and eggs with an electric mixer on medium speed until just combined.

4. Stir in the flour mixture until just combined. Add the sour milk and combine well. Scrape the sides and bottom of the bowl and beat for 1 minute more.

5. Pour the batter into the prepared pan and bake for 50 to 55 minutes, until the edges of the cake pull away from the pan.

6. Remove from the oven and let cool on a rack before serving.

Tip: You can divide this recipe to fill two 8- to 9-inch round cake pans, or you can pour the batter into muffin tins to make cupcakes. This cake is delicious served warm with a drizzle of honey and fresh cream or whole milk from your home dairy.

REFERENCES

American Public Gardens Association. "Frost/Freeze Data." publicgardens.org/resources/frostfreeze-data.

Center for Agriculture, Food and the Environment, University of Massachusetts-Amherst. "Housing and Working Facilities for Goats." https://ag.umass.edu/crops-dairy-livestock-equine/fact-sheets/housing-working-facilites-for-goats.

Mercia, Leonard S. *Storey's Guide to Raising Poultry*. North Adams, MA: Storey Publishing, 2001.

Michigan State University Extension. "Rabbit Tracks: Feeds and Feeding." April 24, 2017. canr.msu.edu/resources/rabbit_tracks_feeds_and_feeding.

National Center for Home Food Preservation. nchfp.uga.edu.

US Department of Agriculture Agricultural Marketing Service. "Retail Milk Prices Report." December 23, 2019. ams.usda.gov/sites/default/files/media/RetailMilkPrices2019.pdf.

US Department of Agriculture Agricultural Research Service. "USDA Plant Hardiness Zone Map." planthardiness.ars.usda.gov.

US Department of Agriculture Cooperative Extension. "How Much Feed Does a Growing Pig Eat a Day?" August 28, 2019. swine.extension.org/how-much-feed-does-a-growing-pig-eat-a-day.

US Department of Agriculture Economic Research Service. "Fruit and Vegetable Prices." ers.usda.gov/data-products/fruit-and-vegetable-prices/fruit-and-vegetable-prices.

US Department of Agriculture National Agricultural Library. "Laws and Regulations." nal.usda.gov/afsic/laws-and-regulations.

US Department of Agriculture National Agricultural Statistics Service. "Crop Production 2018 Summary." February 2019. nass.usda.gov/Publications/Todays_Reports/reports/cropan19.pdf.

———. "Noncitrus Fruits and Nuts 2018 Summary." June 2019. nass.usda.gov/Publications/Todays_Reports/reports/ncit0619.pdf.

———. "Vegetables 2018 Summary." March 2019. downloads.usda.library.cornell.edu/usda-esmis/files/02870v86p/gm80j322z/5138jn50j/vegean19.pdf.

———. "Vegetables 2019 Summary." February 2020. downloads.usda.library.cornell.edu/usda-esmis/files/02870v86p/0r967m63g/sn00bf58x/vegean20.pdf.

www.ingramcontent.com/pod-product-compliance
Lightning Source LLC
Chambersburg PA
CBHW080517030426

42337CB00023B/4553